Serge Lang

ABELIAN VARIETIES

Springer-Verlag
New York Berlin Heidelberg Tokyo

Serge Lang
Yale University
Department of Mathematics
Box 2155 Yale Station
New Haven, Connecticut 06520
U.S.A.

AMS Subject Classification: 14KXX

Library of Congress Cataloging in Publication Data
Lang, Serge, 1927–
Abelian varieties.
Bibliography: p.
Includes index.
1. Abelian varieties. I. Title.
QA564. L277 1983 512'.55 83-10373

Published in 1983 by Springer-Verlag New York Inc.;
originally published in 1959 by Interscience Publishers, Inc.

Printed and bound by R. R. Donnelley & Sons, Harrisonburg, VA.
Printed in the United States of America.

9 8 7 6 5 4 3 2 1

ISBN 0-387-90875-7 Springer-Verlag New York Berlin Heidelberg Tokyo
ISBN 3-540-90875-7 Springer-Verlag Berlin Heidelberg New York Tokyo

PREFACE TO THE SPRINGER EDITION

Abelian Varieties has been out of print for a while. Since it was written, the subject has made some great advances, and Mumford's book giving a scheme theoretic treatment has appeared (D. Mumford, *Abelian Varieties*, Tata Lecture Notes, Oxford University Press, London, 1970). However, some topics covered in my book were not covered in Mumford's; for instance, the construction of the Picard variety, the Albanese variety, some formulas concerning numerical questions, the reciprocity law for correspondences and its application to Kummer theory, Chow's theory for the K/k-trace and image, and others. Several people have told me they still found a number of sections of my book useful. Therefore I thank Springer-Verlag for the opportunity to keep the book in print.

S. Lang

FOREWORD

Pour des simplifications plus substantielles, le développement futur de la géometrie algébrique ne saurait manquer sans doute d'en faire apparaitre.

It is with considerable pleasure that we have seen in recent years the simplifications expected by Weil realize themselves, and it has seemed timely to incorporate them into a new book.

We treat exclusively abelian varieties, and do not pretend to write a treatise on algebraic groups. Hence we have summarized in a first chapter all the general results on algebraic groups that are used in the sequel. They are all foundational results.

We then deal with the Jacobian variety of a curve, the Albanese variety of an arbitrary variety, and its Picard variety, i.e., the theory of cycles of dimension 0 and codimension 1. As we shall see, the numerical theory which gives the number of points of finite order on an abelian variety, and the properties of the trace of an endomorphism are simple formal consequences of the theory of the Picard variety and of numerical equivalence. The same thing holds for the Lefschetz fixed point formula for a curve, and hence for the Riemann hypothesis for curves.

Roughly speaking, it can be said that the theory of the Albanese and Picard variety incorporates in purely algebraic terms the theory which in the classical case would be that of the first homology group. It is far from giving a complete theory of abelian varieties, and a partial list of topics which we do not discuss includes the following:

The theory of differential forms and the cohomology theory.

The infinitesimal and global theory proper to characteristic p.

The theory of linear systems and the Riemann-Roch theorem.

The theory of moduli, i.e., the classification of algebraic families of abelian varieties, and the characterization of Jacobians among abelian varieties.

Various applications to arbitrary varieties, such as, for instance, the equivalence criteria and the theorem of Néron-Severi.

Arithmetic applications like class field theory (which actually belongs to the general theory of algebraic groups) or the theorem of Mordell-Weil.

To a large extent, these topics have not reached the same state of maturity as those with which we deal in this book. Many deserve to have a whole book devoted to them. In any case, we have included at least all the results of Weil's treatise [85] (and, of course, considerably many more).

We shall now make some remarks concerning the formal structure of the book. We begin by a list of prerequisites necessary for a rigorous understanding of the proofs given here. It should be understood, however, that much less is actually required for a general appreciation of the results stated and the methods of proofs. We hope that a good acquaintance with the language of algebraic geometry would suffice.

At the end of each chapter, we append a historical and bibliographical notice, one of whose purposes is to acquaint the reader with the current literature. We have also made comments concerning some of the directions in which the present research is leading. Further historical comments of a more general nature have been made preceding the bibliography given at the end of the book. The index includes all the terms defined here, and the table of notation includes the symbols used most frequently. Finally, we point out that the reader who wishes to get a more detailed summary account of the contents of this book can get it by reading through the brief introductions with which we begin each chapter.

S. Lang

New York, Fall 1958

PREREQUISITES

They are of order 4.

1. Elementary qualitative algebraic geometry, as it is treated for instance in *Introduction to Algebraic Geometry*, Interscience, New York, 1958. This book will be referred to as IAG. It treats of varieties, cycles, linear systems, topics in field theory, Zariski topology, and other topics of a heterogeneous nature.

2. The Riemann-Roch theorem for curves.

3. The elementary theory of algebraic groups: definitions, subgroups, factor groups, and the possibility of recovering a group starting with birational data. We have recalled all the results needed in Chapter I, without proofs. A complete self-contained exposition can be found in [90], [91], [92].

4. Intersection theory, of type $F\text{-}X_y$ Th. Z. Occasionally we have given an argument in the language of specialization of cycles, for which we refer to Matsusaka [55].

In an appendix we have recalled certain theorems on correspondences properly belonging to the *Foundations of Algebraic Geometry*.

The terminology is that of *Foundations* [83] except for the following modifications.

Let f: $U \to V$ be a rational map. We say that f is *defined over a field k* if k is a field of definition for U, V and the graph of f, usually denoted by Γ_f. Let P be a point of U. We say that f is *holomorphic* at P, or *defined at P*, instead of saying (as in *Foundations*) that f is regular at P.

On the other hand, let f be defined over k, and let u be a generic point of U over k. Let $v = f(u)$. We shall say that f is respectively *regular, separable, primary, purely inseparable* if the extension $k(u)$ of $k(v)$ is of the corresponding type. If v is a generic point of V over k, we say that f is *generically surjective*. Suppose this is the case. Then one sees easily that the above four conditions are respectively equivalent to the following ones concerning the cycle

$$f^{-1}(v) = \mathrm{pr}_1 \, [\Gamma_f \cdot (U \times v)]:$$

It is a variety with multiplicity 1.

All the components of $f^{-1}(v)$ have multiplicity 1.

There is only one component having multiplicity p^m (where p is the characteristic).

It has only one component, which is a point with multiplicity p^m.

We observe that in the above notation, the support of $f^{-1}(v)$ is the locus of u over $k(v)$.

Let $f\colon U \to V$ be again a rational map of U into V, defined over k. Let W be a subvariety of U also defined over k. Let w be a generic point of W over k. We say that f is *defined at* W if f is defined at w. The locus of the point $f(w)$ over k will then be denoted by $f(W)$. It is in general distinct from the cycle $\mathrm{pr}_2 \, [\Gamma_f \cdot (W \times V)]$ even when this intersection is defined. We shall also use $f(W)$ to denote this cycle, and the context will usually make our meaning clear. To avoid confusion, we may also call the first the *set-theoretic image* $f(W)$, and the second the *cycle* $f(W)$, or $f(W)$ *in the sense of intersection theory*.

Finally (added in proof), to conform with the functorial terminology which is generally becoming accepted, we would like to recommend the use of the word "isomorphism" instead of the words "birational isomorphism." What we here call "isomorphism" should be called a "bijective homomorphism."

CONTENTS

CHAPTER I

Algebraic Groups

1. Groups, subgroups, and factor groups 1
2. Intersections and Pontrjagin products 6
3. The field of definition of a group variety 13

CHAPTER II

General Theorems on Abelian Varieties

1. Rational maps of varieties into abelian varieties 20
2. The Jacobian variety of a curve 30
3. The Albanese variety 40

CHAPTER III

The Theorem of the Square

1. Algebraic equivalence 55
2. The theorem of the cube and the theorem of the square . . 67
3. The theorem of the square for groups 71
4. The kernel in the theorem of the square 76

CHAPTER IV

Divisor Classes on an Abelian Variety

1. Applications of the theorem of the square to abelian varieties 86
2. The torsion group 94
3. Numerical equivalence 101
4. The Picard variety of an abelian variety. 114

CHAPTER V

Functorial Formulas

1. The transpose of a homomorphism 123
2. A list of formulas and commutative diagrams 126
3. The involutions 132

CHAPTER VI

The Picard Variety of an Arbitrary Variety

1. Construction of the Picard variety 147
2. Divisorial correspondences 153
3. Application to the theory of curves 155
4. Reciprocity and correspondences 165

CHAPTER VII

The l-Adic Representations

1. The l-adic spaces 179
2. Dual representations 187

CHAPTER VIII

Algebraic Systems of Abelian Varieties

1. The K/k-image 198
2. The generic hyperplane section 208
3. The K/k-trace 211
4. The transpose of an exact sequence 216
5. Duality between image and trace 222
6. Exact sequences of varieties 224

APPENDIX

Composition of Correspondences

1. Inverse images 231
2. Divisorial correspondences 238

Bibliography 245

Table of Notation 253

Index 255

CHAPTER I

Algebraic Groups

The purpose of this chapter is to recall briefly the fundamental notions of the theory of algebraic groups. In the sequel, we shall use only elementary properties of algebraic groups, and we shall not need structure theorems, for instance. All the results which we shall need are stated explicitly below. We give no proofs in § 1. Granting IAG, a complete self-contained exposition can be found in the papers of Weil and Rosenlicht.

The numerical theory in § 2, together with the Pontrjagin products requires intersection theory, and the proofs depend on *Foundations*.

Finally in § 3, we have stated the theorems concerning the field of definition of a variety, and indicated how they can be used to lower the field of definition of a group variety provided certain coherent isomorphisms are given. For the proof, we refer the reader to [92]. We shall use § 3 in the sequel only at the end of the theory of the Albanese variety, and in the last chapter, for the theory of algebraic systems of abelian varieties. The rest of this book is independent of § 3, and we advise the reader to skip § 2 and § 3 until he comes to a place where they are used.

§ 1. *Groups, subgroups, and factor groups*

An *algebraic group* is the union of a finite number of disjoint varieties (abstract) G_α called its components, on which a group structure is given by everywhere defined rational maps. More precisely, for each pair G_α, G_β of components of G, we are given a rational map $f_{\alpha\beta} : G_\alpha \times G_\beta \to G_\gamma$, everywhere defined into a third component G_γ determined by α and β, such that the group law $(x, y) \to xy$ for $x \in G_\alpha$ and $y \in G_\beta$ is given by $xy = f_{\alpha\beta}(x, y)$. In addition, for each G_α, we are given a birational biholomorphic map $\varphi_\alpha : G_\alpha \to G_\lambda$ into another component G_λ such that the inverse $x \to x^{-1}$ is given by $x^{-1} = \varphi_\alpha(x)$.

If there is only one component, the algebraic group is called a *group variety*, or a connected algebraic group. An algebraic group is defined over a field k if all the G_α are defined over k and the $f_{\alpha\beta}$, φ_α are also defined over k. One then sees that the identity e of G is rational over k.

Let G be a group variety. Then G is non-singular. This comes from the fact that for each point $a \in G$ there is a birational biholomorphic transformation $T_a : G \to G$ of G onto itself such that $T_a(x) = ax$. We call it the *left translation*. On the whole, we shall deal only with commutative groups and thus do not need to distinguish our left from our right. If U is a subvariety of G, we denote by aU, or U_a, the left translation of U by a, i.e., $T_a(U)$.

Let H be an abstract subgroup of the group variety G, and assume that H is also an algebraic subset of G. Then H is an algebraic group, whose law of composition is induced by that of G. The components of H are all translations of the connected component (of identity) of H. If G is defined over k, and if H is k-closed, then the connected component of H is also k-closed, and is therefore defined over a purely inseparable extension of k. Indeed, every automorphism of the universal domain leaving k fixed leaves H fixed, permutes the components of H, and must leave the connected component of identity fixed because it leaves e fixed since e is rational over k.

Let G, G' be group varieties.

By a *rational homomorphism*, or simply *homomorphism* $\lambda : G \to G'$ we shall mean an everywhere defined rational map of G into G' which verifies the condition $\lambda(xy) = \lambda(x)\lambda(y)$, i.e., is an abstract homomorphism. We shall say that λ is an *isomorphism* if it is injective (i.e., one-one). As a rational map, λ is then purely inseparable, of finite degree. If this degree is equal to 1, we shall call λ a *birational isomorphism*.

We say that λ is *defined over* k if G, G', and the graph of λ are defined over k. This is compatible with our definition for rational maps.

Let $\lambda : G \to G'$ be a homomorphism defined over k. Then λ is continuous (for the k-topology of Zariski, of course) and its kernel is therefore an algebraic subgroup H of G, which is k-closed.

Let H be an algebraic subgroup of a group variety G and assume that H is a normal subgroup (in the abstract sense). *Then one can give the factor group the structure of a group variety.* More precisely, *there exists a group variety G' and a surjective homomorphism $\lambda : G \to G'$ such that:*

(i) *the kernel of λ is equal to H;*

(ii) *the map λ is separable;*

(iii) *the pair (G', λ) satisfies the universal mapping property for homomorphisms of G whose kernel contains H.*

More precisely, if $\alpha : G \to G''$ is a homomorphism of G whose kernel contains H, then there exists a homomorphism $\beta : G' \to G''$ such that $\alpha = \beta\lambda$. We shall call λ the *canonical homomorphism* on the factor group G'.

We shall say that the algebraic subgroup H of G is *rational* over k if the cycle consisting of the components of H taken with multiplicity 1 is rational over k. If G is defined over k, and if H is rational over k, then we may take the canonical homomorphism $\lambda : G \to G' = G/H$ also defined over k.

Suppose in addition that H is connected and both G, H are defined over k. Let b be a point of G'. Then $\lambda^{-1}(b) = H_a$ for any point $a \in G$ such that $\lambda(a) = b$. This is true both set-theoretically and in the sense of intersection theory i.e., $\lambda^{-1}(b)$ has exactly one component with multiplicity 1. Let u be a generic point of G over k, and put $v = \lambda(u)$. Then $\lambda^{-1}(v) = H_u$. Furthermore, H_u is a *homogeneous space* for H under left translation. The variety H_u is defined over $k(v)$, and is the locus of u over $k(v)$ according to the general theory of rational maps.

More generally, we shall say that a variety V is a *homogeneous space* of a group variety G if we are given an everywhere defined rational map $f : G \times V \to V$ such that if we write xP instead of $f(x, P)$, then $(xy)P = x(yP)$, and for any two points P, Q of V, there exists an element $x \in G$ such that $xP = Q$. In particular, if P is a generic point of V over a field of definition k for f, and if x is a generic point of G over $k(P)$, then xP is a generic point of V over $k(P)$. We shall almost never use homogeneous spaces in the sequel. The only point where a homogeneous space will occur will be in the proof of the complete reducibility theorem of Poincaré.

For the sake of completeness, recall that a homogeneous space V is said to be *principal* if the operation of G is simply transitive and is separable. In other words, if P is a point of V, and x is a generic point of G over $k(P)$ then the map $x \to xP$ establishes a birational biholomorphic correspondence between G and V. A homogeneous space V defined over k may of course not have a rational point over k. The search for conditions under which it has such points is an interesting diophantine problem.

In the example of the factor group above, the coset H_u is in fact a principal homogeneous space for H, defined over $k(v)$.

One can recover a group variety from birational data in the following manner.

Let V be an arbitrary variety, and suppose we are given a *normal law of composition*. By this we mean a rational map $f: V \times V \to V$ which is generically surjective, and such that if u, v are two independent generic points of V over a field of definition k for f, then $w = f(u, v)$ is a generic point of V over k, and $k(u, v) = k(v, w) = k(u, w)$. In addition, f is assumed to be generically associative, i.e., if u, v, w are three independent generic points of V over k, then

$$f(u, f(v, w)) = f(f(u, v), w).$$

We denote $f(u, v)$ by uv. If U is a variety birationally equivalent to V, and if $T: V \to U$ is a birational transformation, then we can obviously define a law of composition on U by the formula

$$T(v)T(u) = T(w).$$

We say that this law is obtained by *transferring* that of V. A fundamental theorem then asserts that *if V and its normal law f are defined over k, there exists a group variety G also defined over*

k, such that the law of composition on G is obtained by transferring that of V. This group G is uniquely determined up to a birational isomorphism.

This uniqueness property is an immediate consequence of the following remarks which are extremely useful in handling group varieties.

Let $\lambda : G \to G'$ be a rational map of a group variety into another one, and assume that λ satisfies

$$\lambda(xy) = \lambda(x)\lambda(y)$$

whenever x, y are independent generic points of G. We shall then say that λ is a *generic homomorphism. It then follows that λ is everywhere defined, is a homomorphism, and that $\lambda(G)$ is a group subvariety of G'*. Indeed, we can write $\lambda(x) = \lambda(xy)\lambda(y)^{-1}$. For any x, we take y generic. Then xy and y are generic, and this shows that λ is defined at x. From this we conclude that λ is a homomorphism. Let H' be the closure of $\lambda(G)$ in G' for the Zariski topology. Then $\lambda(G)$ contains a non-empty open subset of H'. Since H' is the closure of an abstract subgroup of G', it is a subgroup of G', and the cosets of $\lambda(G)$ contain a non-empty open subset of H'. This can happen only if $\lambda(G) = H'$.

We finish this paragraph by stating an important property of commutative groups.

Let G be a commutative group variety. Let $\mathfrak{a}$ be a cycle of dimension 0 on G. It is a formal sum of points, which we write

$$\mathfrak{a} = \sum n_i(x_i).$$

Writing the law of composition on G additively, we can take the sum of the points x_i on G, each one taken n_i times. We thus obtain a point of G which will be denoted by $S(\mathfrak{a})$. This sum will be written without parentheses, to distinguish it from the formal sum above. Thus we have

$$S(\mathfrak{a}) = \sum n_i \cdot x_i = \sum n_i x_i.$$

In this notation, if x, y are two points of G, then $(x) + (y)$ is the 0-cycle of degree 2 having x, y as components with multiplicity 1, while $x + y$ is the sum on G of x and y.

Let k be a field of definition for G. The *fundamental theorem on symmetric functions* then asserts that *if* $\mathfrak{a}$ *is a* 0-*cycle of* G, *rational over* k, *then the point* $S(\mathfrak{a})$ *is rational over* k.

Of course, this is a special case of a more general theorem concerning arbitrary symmetric functions ([85] Th. 1), but the above statement will suffice for this book. It is obvious in the case where all the points of $\mathfrak{a}$ are rational over a separable extension of k. Indeed, the composition law of G being defined over k, for every automorphism σ of the algebraic closure of k leaving k fixed, we have

$$(S(\mathfrak{a}))^{\sigma} = S(\mathfrak{a}^{\sigma}) = S(\mathfrak{a}).$$

This shows that in general, $S(\mathfrak{a})$ is purely inseparable over k. The proof in this case is pure technique in characteristic p.

Left $f : U \to G$ be a rational map of a variety into the commutative group variety G. Let $\mathfrak{a}$ be a 0-cycle on U, and assume that f is defined at all the points of $\mathfrak{a}$. Then $f(\mathfrak{a})$ is a cycle on G: If $\mathfrak{a} = \sum n_i(P_i)$, then $f(\mathfrak{a}) = \sum n_i(f(P_i))$. Furthermore, if f is defined over k, and if $\mathfrak{a}$ is rational over k, then $f(\mathfrak{a})$ is rational over k, because

$$f(\mathfrak{a}) = \mathrm{pr}_2[\Gamma_f \cdot (\mathfrak{a} \times G)].$$

It follows that $S(f(\mathfrak{a}))$ is a point of G, rational over k. We shall denote it by $S_f(\mathfrak{a})$.

§ 2. *Intersections and Pontrjagin products*

We shall give here special formulas concerning intersections on group varieties. They show how certain operations can be defined in terms of intersection theory.

PROPOSITION 1. *Let* G *be a group variety,* V *a subvariety of* G, *both defined over* k. *Let* (u, x) *be independent generic points of* G, V *over* k. *Let* $\bar{V}$ *be the locus of* (u, ux) *over* k, ($(u, u+x)$ *if* G *is commutative*). *Then we have*

$$\bar{V} \cdot (u \times G) = u \times V_u.$$

Proof: We need but to apply F—VII$_6$ Th. 12.

We shall denote by $s_n : G \times \ldots \times G \to G$ the rational map of the product of G with itself n times obtained by the formula $s_n(u_1, \ldots, u_n) = u_1 \ldots u_n$. If G is commutative, then s_n is a homomorphism, which will be called the *sum*. In the non-commutative case, we say it is the *product*. Its graph will be denoted by S_n.

PROPOSITION 2. *Let $s_2 : G \times G \to G$ be the product, and V a subvariety of G. Let (u, x) be as in Proposition* 1. *Then the cycle*

$$s_2^{-1}(V) = \mathrm{pr}_{12}[S_2 \cdot (G \times G \times V)]$$

is a variety, which is the locus of $(u, u^{-1}x)$ over k, and we have

$$s_2^{-1}(V) \cdot (u \times G) = u \times u^{-1}V$$

or in the additive case, $u \times V_{-u}$.

Proof: Every point (a, b, c) of $S_2 \cap (G \times G \times V)$ is such that $ab = c$ and c is in V. We see therefore that the support of $s_2^{-1}(V)$ is a variety, locus of the point $(u, u^{-1}x)$. The single component of $S_2 \cdot (G \times G \times V)$ has multiplicity 1, according to F—VII$_6$ Th. 17. Since s_2 is everywhere defined on $G \times G$, the projection on the first two factors conserves this multiplicity.

Note particularly the sign $-$ in the intersection

$$s_2^{-1}(V) \cdot (u \times G) = u \times V_{-u}.$$

We shall now define the Pontrjagin products. Let V, W be two subvarieties of G. We denote by $V \otimes W$ their set-theoretic product on G, or if G is commutative, by $V \oplus W$ or $V + W$. If x, y are two independent generic points of V, W over a field k, then by definition, $V \otimes W$ is the locus of xy over k. We have a rational map $F : V \times W \to V \otimes W$ induced by s_2, and we shall denote the degree of F by $d(V, W)$ if it is finite, and by 0 otherwise. We thus have $d(V, W) = \nu(F)$.

PROPOSITION 3. *Let V, W be two subvarieties of G. Then*

$$\mathrm{pr}_3[S_2 \cdot (V \times W \times G)] = d(V, W)(V \otimes W).$$

Proof: Since s_2 is everywhere defined, $S_2 \cdot (V \times W \times G)$ has one component with multiplicity 1 by F—VII$_6$ Th. 17. Our proposition is then a consequence of the definition of the projection.

The cycle $d(V, W)(V \otimes W)$ will be denoted by $V * W$. We have $V * W = 0$ if and only if the dimension of $V \otimes W$ is smaller than $\dim V + \dim W$. We shall say that $V \otimes W$, or $V * W$, is the *Pontrjagin product* of V and W. It will always be clear from the context whether we mean the set theoretic product, or the cycle.

If V is a point $a \in G$, then $V \otimes W = W_a$ is the translation of W by a.

The Pontrjagin product is associative. In order to see this, let U, V, W be three subvarieties of G, defined over k, and let x, y, z be three independent generic points of U, V, W over k. It is clear that $(U \otimes V) \otimes W = U \otimes (V \otimes W)$, this variety being the locus of xyz over k. On the other hand, if we put $d = d(U, V)$ and $e = d(U \otimes V, W)$, then $de = [k(x, y, z) : k(xyz)]$ if this degree is finite, and 0 otherwise.

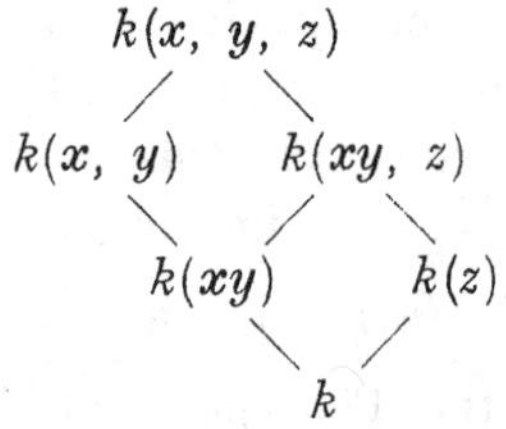

Indeed, $k(x, y)$ and $k(z)$ are linearly disjoint over k, and hence $[k(x, y) : k(xy)] = [k(x, y, z) : k(xy, z)]$. Our assertion is then obvious, taking into account the inclusion

$$k(xyz) \subset k(xy, z) \subset k(x, y, z).$$

We have thus shown that

$$U * (V * W) = (U * V) * W,$$

and that the Pontrjagin product is associative from the point of view of intersection theory.

Of course, we can define the symbol $d(V_1, \ldots, V_m)$ for several subvarieties of G: It is the degree of the rational map of the product $V_1 \times V_2 \times \ldots \times V_m$ on the Pontrjagin product $V_1 \otimes V_2 \otimes \ldots \otimes V_m$ if it is finite, and 0 otherwise. We then have

$$V_1 * V_2 * \ldots * V_m = d(V_1, \ldots, V_m)(V_1 \otimes V_2 \otimes \ldots \otimes V_m).$$

By linearity, we may extend the Pontrjagin product to cycles, and for two cycles Z_1, Z_2 on G we have clearly

$$Z_1 * Z_2 = \mathrm{pr}_3[S_2 \cdot (Z_1 \times Z_2 \times G)].$$

Thus the Pontrjagin product defines a ring structure on the cycles of G.

If G is commutative, then the Pontrjagin product is also commutative.

Let now G be arbitrary, and let $\alpha : G \to G'$ be a homomorphism. If V is a subvariety of G, then $\alpha(V)$ in the sense of intersection theory is the cycle

$$\mathrm{pr}_2[\Gamma_\alpha \cdot (V \times G')].$$

This cycle has exactly one component V', which appears with multiplicity $[V : V']$ if this degree is finite, and 0 otherwise. We shall now show that α induces a homomorphism of the Pontrjagin ring of G into that of G', i.e., that we have

$$\alpha(V_1 * V_2 * \ldots * V_m) = \alpha(V_1) * \alpha(V_2) * \ldots * \alpha(V_m).$$

Since the product is associative, it suffices to show this for two factors V, W, i.e., we must show

$$\alpha(V * W) = \alpha(V) * \alpha(W).$$

Let k be a field of definition for α, V, and W. Let x, y be two independent generic points of V, W over k, put $x' = \alpha(x)$, $y' = \alpha(y)$ and let V', W' be the loci of x' and y', respectively, over k so that V' and W' are the images of V and W by α in the set theoretic sense.

Consider the following field inclusions:

$$k(x, y) \supset k(xy) \supset k(x'y')$$
$$k(x, y) \supset k(x', y') \supset k(x'y').$$

If the extension $k(x, y)$ of $k(x'y')$ is not finite, then one sees immediately that both sides of our formula are equal to 0. Assume the extension finite. From the definitions, we see that the multiplicity of the single component in the left-hand side of our formula is then equal to

$$[k(x, y) : k(xy)][k(xy) : k(x'y')] = [k(x, y) : k(x'y')].$$

On the other hand, since $k(x)$ and $k(y)$ are linearly disjoint over k, we have

$$[k(x, y) : k(x', y')] = [k(x) : k(x')][k(y) : k(y')].$$

By definition, the multiplicity of the component on the right-hand side of the formula is therefore equal to

$$[k(x, y) : k(x', y')][k(x', y') : k(x'y')] = [k(x, y) : k(x'y')].$$

This proves our formula.

We shall now study more closely an important special case of Pontrjagin products, namely the case where V, W are of complementary dimension.

Observe first that if G, V, W are defined over k, and if u is a generic point of G over k, then the intersection $V \cdot W_u$ is defined. In order to show this, let us denote by W^- the variety obtained from W by the transformation $x \to x^{-1}$ of G. Without loss of generality, we may consider $V \cdot W^-_u$ instead of $V \cdot W_u$. Let x be a point of $V \cdot W^-_u$. It is *a fortiori* a point of V, and we can write $x = uy^{-1}$ with $y \in W$, whence $xy = u$. Since $\dim_k(x) + \dim_k(y) = \dim G$, and since $\dim_k(x) \leqq \dim V$ and $\dim_k(y) \leqq \dim W$, we see that x and y are independent generic points of V, W over k, and that x, y are algebraic over $k(u)$, whence $V \cdot W^-_u$ is defined.

We contend that the points of intersection of $V \cdot W^-_u$ consist precisely of the conjugates of x over $k(u)$. Indeed, if σ is an isomorphism of $k(x, y)$ over $k(u)$, then $x^\sigma y^\sigma = u$, and hence every conjugate of x over $k(u)$ is a point of the intersection. Conversely, if $x_1 \in V$ and $y_1 \in W$ are such that $x_1 y_1 = u$, then from what we have seen above they are independent generic points of V, W over k, and hence there exists an isomorphism $\sigma : k(x, y) \to k(x_1, y_1)$ such that $x_1 = x^\sigma$ and $y_1 = y^\sigma$. We see that $u = (x_1 y_1) =$

$x^\sigma y^\sigma = (xy)^\sigma = u^\sigma$ is invariant under σ, and hence we conclude that x_1 is conjugate to x over $k(u)$.

We shall see below that the multiplicity of each point is equal to the inseparability degree of $k(x, y)$ over $k(u)$, and therefore that our intersection is a prime rational cycle over $k(u)$.

PROPOSITION 4. *Let X, Y be two cycles on G and let a be a point of G such that $X \cdot Y_a$ is defined. Then*

$$a \times X_a \cdot Y = \mathrm{pr}_{13}\,[S_2 \cdot (a \times X \times Y)] = Z \cdot (a \times G)$$

where Z is the cycle obtained from $X \times Y$ by the transformation of $G \times G$ which maps each point (x, y) onto (yx^{-1}, y). If a is a point of G such that $X \cdot Y^-{}_a$ is defined, then we have

$$a \times Y^-{}_a \cdot X = \mathrm{pr}_{13}\,[S_2 \cdot (a \times Y^- \times X)] = Z_1 \cdot (a \times G)$$

where Z_1 is the cycle obtained from $X \times Y$ by the transformation of $G \times G$ which maps each point (x, y) onto (xy, x).

Proof: We have included the second formula for the convenience of the reader: for the proof we may obviously restrict ourselves to the first. Moreover, by linearity, we may restrict ourselves to the case where X, Y are varieties. Let X' be the subvariety of $G \times G$ consisting of all pairs of points (x, ax) with $x \in X$. We have

$$S_2 \cdot (a \times X \times G) = a \times X'.$$

If we apply the associativity theorem to the intersection

$$S_2 \cdot (a \times X \times G) \cdot (G \times G \times Y)$$

we find

$$a \times [X' \cdot (G \times Y)] = S_2 \cdot (a \times X \times Y).$$

Taking pr_{13} and using F—VII_6 Th. 17, Cor. 1 on the left-hand side, we obtain

$$a \times (X_a \cdot Y) = \mathrm{pr}_{13}[S_2 \cdot (a \times X \times Y)]$$

thus proving the first part of our proposition.

As to the second, we consider the intersection

$$S_2 \cdot (G \times X \times Y).$$

From right to left, we may apply F—VII_6 Th. 17. Since the projection on the last factor is holomorphic, this intersection has one component, having the same multiplicity as

$$(\mathrm{pr}_{23}S_2) \cdot (X \times Y) = (G \times G) \cdot (X \times Y),$$

which is therefore equal to 1. Our component is the locus of (yx^{-1}, x, y) with $x \in X$ and $y \in Y$. We thus obtain

$$\mathrm{pr}_{13}[S_2 \cdot (G \times X \times Y)] = Z$$

which is the locus of (yx^{-1}, y). But

$$S_2 \cdot (a \times X \times Y) = S_2 \cdot (G \times X \times Y) \cdot (a \times G \times G).$$

If we take pr_{13} on both sides, we get

$$a \times (X_a \cdot Y) = Z \cdot (a \times G),$$

thus proving our proposition.

If we apply F—VII_6 Th. 12 to the intersection $Z \cdot (u \times G)$ with u generic on G, we obtain the result mentioned above, namely:

PROPOSITION 5. *Let V, W be two subvarieties of G, of complementary dimension, and let u be a generic point of G over a field of definition k for G, V, W. Then the cycle $V \cdot W^{-}{}_{u}$ is defined, and consists of those points $x \in V$ for which there exists $y \in W$ such that $xy = u$. Each one of those points appears with a multiplicity equal to the inseparability degree of $k(x, y)$ over $k(u)$. They form a complete system of conjugates over $k(u)$, and we have*

$$\deg (V \cdot W^{-}{}_{u}) = d(V, W) = [k(x, y) : k(u)].$$

If V, W are of complementary dimension, then

$$V * W = d(V, W)G$$

and $d(V, W) = 0$ if and only if the dimension of $V \otimes W$ is smaller than that of G. We have $V \cdot W^{-}{}_{u} = 0$ if and only if $V * W = 0$. The cycles of maximal dimension on G are all integral multiples of G, and if a cycle is of type $m \cdot G$ we shall say that m is its *degree*. With this notation, we then have the following result which follows immediately from Proposition 5.

PROPOSITION 6. *Let X, Y be two cycles on G of complementary dimension, rational over a field of definition k for G, and let u be a generic point of G over k. Then*

$$\deg (X \cdot Y^{-}{}_{u}) = d(X, Y) = \deg (X * Y).$$

We conclude this section with a result which is of considerable importance, because it gives in a natural way the possibility of defining the intersection of two arbitrary algebraic systems of cycles on a group.

PROPOSITION 7. *Let G be a group variety, V an arbitrary variety, X a cycle on G, and Z a cycle on $G \times V$. Let k be a field of definition for G, V over which X and Z are rational. Let u be a generic point of G over k. Then the intersection $Z \cdot (X_u \times V)$ is defined.*

Proof: Let $\dim G = n$, $\dim X = r$, and $\dim Z = s$. To prove our proposition, we must show that the dimension of a component of $Z \cap (X_u \times V)$ is $\leqq r + s - n$. Without loss of generality, we may assume that X, Z are varieties. Let (ux', M') be a point in $Z \cap (X_u \times V)$. We must show that the dimension of this point over $k(u)$ is $\leqq r + s - n$. Since $(ux', M') \in Z$, we must have $\dim_k (ux', M') \leqq s$. Since $x' \in X$, we must have $\dim_k (x') \leqq r$. Hence $\dim_k (x', ux', M') \leqq r + s$. But the field $k(x', ux', M')$ contains $k(u)$, which has dimension n over k. Hence the dimension of (u', ux', M') over $k(u)$ is at most $r + s - n$. Since the field generated over $k(u)$ by (x', ux', M') is the same as the field generated by (ux', M'), we have proved our proposition.

§ 3. The field of definition of a group variety

Let V be a variety defined over a field K containing a field k. We are going to recall certain criteria which allow us to obtain a variety V_0 defined over k and a birational correspondence $f : V_0 \to V$ defined over K. We shall then consider the special case where V is a group variety.

We begin by considering a regular extension of k, which we may assume finitely generated. Let T be a variety defined over k, which we consider as a parameter variety, and let t be a generic point of T over k. We denote by V_t a variety defined over $k(t)$, and if t' is another generic point of T over k, we denote by $V_{t'}$ the transform of V_t by the isomorphism of $k(t)$ which maps t on t'. Similarly, if f_t is a rational map defined over $k(t)$, then $f_{t'}$ will be its transform by this isomorphism. If t, t', t'' are three independent generic points of T over k, and $f_{t',t}$ is a rational map defined over $k(t, t')$, we denote by $f_{t'',t'}$ its transform by the isomorphism of $k(t, t')$ which maps (t, t') on (t', t'').

In the following statement, $k(t)$ is a regular extension of k, and t is a generic point of a parameter variety T.

THEOREM 1. *Let V_t be a variety defined over a regular extension $k(t)$ of k, and assume that there exists a birational correspondence*

$$f_{t',t} : V_t \to V_{t'}$$

defined over $k(t, t')$, satisfying the coherence condition

$$f_{t'',t} = f_{t'',t'} f_{t',t}.$$

Then there exists a variety V defined over k, and a birational transformation

$$f_t : V \to V_t$$

defined over $k(t)$ such that $f_{t',t} = f_{t'} f_t^{-1}$, or in other words, $f_{t'} = f_{t',t} f_t$.

We see immediately that V is uniquely determined up to a birational transformation defined over k.

We apply this result to the case where $V_t = G_t$ is a group variety, and we assume in addition that $f_{t',t}$ is a birational isomorphism. In particular, the following condition is satisfied for independent generic points u, v of G_t over $k(t)$:

$$f_{t',t}(uv) = f_{t',t}(u)\, f_{t',t}(v). \tag{1}$$

We note that uv is the product of u, v on G, while the product in the expression on the right-hand side of (1) is taken on $G_{t'}$. If L_t is the graph of the law of composition of G_t, then $L_{t'}$ is the graph of the law of composition on $G_{t'}$.

We can define a law of composition L on V by transfer over the field $k(t)$. If x, y are two independent generic points of V over

$k(t)$, we define

$$L(x, y) = f_t^{-1}[f_t(x)f_t(y)]$$

the product being taken on G_t. If we apply the isomorphism $\sigma : k(t) \to k(t')$ to this expression (having extended σ to the universal domain in such a way that it leaves x, y fixed), and if we use (1) and the relation $f_{t'} = f_{t',t} f_t$ we get

$$L^\sigma(x, y) = L(x, y).$$

Hence L is invariant under σ, and is therefore defined over $k(t) \cap k(t') = k$. A fundamental theorem recalled in § 1 allows us to take for V a group variety G. In addition, we have the relation

$$f_t L(x, y) = f_t(x) f_t(y)$$

which shows that f_t is a generic homomorphism. Since f_t is a birational map, we conclude that f_t is a birational isomorphism.

For some applications, it is useful to consider a somewhat more general situation, where we are given a variety U and a rational map $\varphi_t : U \to G_t$ defined over $k(t)$, satisfying the condition

$$\varphi_{t'} = f_{t',t}\, \varphi_{t}.$$

If we define $\varphi : U \to G$ by the formula $\varphi = f_t^{-1} \varphi_t$, we obtain a rational map which is defined over k. Indeed, the isomorphism of $k(t)$ sending t on t' transforms φ into $f_{t'}^{-1} \varphi_{t'}$, which is equal to φ, taking into account the coherence condition satisfied by φ_t. We have therefore obtained the following complement to Theorem 1.

THEOREM 1G. *Let G_t be a group variety defined over a regular extension $k(t)$ of a field k, and assume that there exists a birational isomorphism $f_{t',t} : G_t \to G_{t'}$ defined over $k(t, t')$, satisfying the coherence condition. Then there exists a group variety G, defined over k, and a birational isomorphism $f_t : G \to G_t$ defined over $k(t)$ such that $f_{t'} = f_{t',t} f_{t'}$. If in addition U is a variety defined over k, and $\varphi_t : U \to G_t$ is a rational map defined over $k(t)$ such that $\varphi_{t'} = f_{t' t}\, \varphi_t$, then there exists a rational map $\varphi : U \to G$ defined over k such that $\varphi_t = f_t \varphi$.*

Let us now consider the case where K is a finite separable algebraic extension of k. For arbitrary varieties, we have the following theorem.

THEOREM 2. *Let k_1 be a finite separable algebraic extension, and let V_1 be a variety defined over k_1. Assume that for each pair (σ, τ) of isomorphisms of k_1 over k, there exists a birational transformation $f_{\tau,\sigma}: V_1^\sigma \to V_1^\tau$ defined over a separable algebraic extension of k, and satisfying the coherence conditions*

(i) $f_{\tau,\rho} = f_{\tau,\sigma} f_{\sigma,\rho}$;

(ii) *for every automorphism ω of the separable closure k_s of k, we have $f_{\tau\omega,\sigma\omega} = (f_{\tau,\sigma})^\omega$.*

Then there exists a variety V defined over k, and a birational transformation $f_1 : V \to V_1$ defined over k_1 such that $f_{\tau,\sigma} = f_1^\tau (f_1^\sigma)^{-1}$.

We see immediately that V is uniquely determined up to a birational transformation defined over k.

We apply Theorem 2 to the case where V_1 is a group variety G_1, and where the $f_{\tau,\sigma}$ are birational isomorphisms. We then have the following condition, for independent generic points u, v of G_1^σ over k_s:

$$f_{\tau,\sigma}(uv) = f_{\tau,\sigma}(u) f_{\tau,\sigma}(v). \tag{2}$$

Again, we can define a law of composition L on V by transfer over the field k_1. If x, y are independent generic points of V over k, we define

$$L(x, y) = f_1^{-1}[f_1(x) f_1(y)]$$

the product being taken on G_1. If we apply any isomorphism σ of k_1 over k, and if we use condition (2) together with the relation $f_{\tau,\sigma} = f_1^\tau (f_1^\sigma)^{-1} = f_1^\tau (f_1^{-1})^\sigma$ we find

$$\begin{aligned} L^\sigma(x, y) &= (f_1^{-1})^\sigma [f_1^\sigma(x) f_1^\sigma(y)] \\ &= (f_1^{-1})^\tau [f_1^\tau(x) f_1^\tau(y)] \\ &= L^\tau(x, y). \end{aligned}$$

If we take τ equal to the identity, we see that L is fixed under all isomorphisms of k_1 over k, and is therefore defined over k.

Finally, if we are given a variety U defined over k, and a rational map $\varphi_1 : U \to G_1$ defined over k_1 satisfying the condition

$$\varphi_1^{\sigma} = f_{\sigma,\tau}\varphi_1^{\tau}$$

for all σ, τ, then we can define $\varphi : U \to G$ by the formula $\varphi = f_1^{-1}\varphi_1$, and we see immediately that $\varphi^{\sigma} = \varphi$, and hence that φ is defined over k. Summarizing the above remarks, we obtain the following complement to Theorem 2.

THEOREM 2G. *Let G_1 be a group variety defined over a finite separable algebraic extension k_1 of k, and assume that for each pair (σ, τ) of isomorphisms of k_1 over k, there exists a birational isomorphism $f_{\tau,\sigma} : G_1^{\sigma} \to G_1^{\tau}$ defined over a separable algebraic extension, and satisfying the coherence conditions* (i) *and* (ii). *Then there exists a group variety G defined over k, and a birational isomorphism $f_1 : G \to G_1$ defined over k_1 such that $f_{\tau,\sigma} = f_1^{\tau}(f_1^{\sigma})^{-1}$. If in addition U is a variety defined over k, and $\varphi_1 : U \to G_1$ is a rational map defined over k_1 such that $\varphi_1^{\sigma} = f_{\sigma,\tau}\varphi_1^{\tau}$, then there exists a rational map $\varphi : U \to G$ defined over k such that $\varphi_1 = f_1\varphi$.*

In practice, we usually have $f_{1,1}$ equal to the identity, and hence $f_{\tau,\sigma} = f_{\sigma,\tau}^{-1}$. We can then write

$$f_1^{\sigma} = f_{\sigma,\tau}f_1^{\tau}$$

instead of the formula in the statement of the theorem.

Historical Note:

Algebraic groups were first defined by Weil [85] who recognized the possibility of recovering a group from birational data. At that time, Weil left two problems open: one concerning the field of definition of the group, and the second concerning the projective embedding.

Chow [17] constructed the Jacobian of a curve by a projective method, and thus obtained in that case the solution of the two problems. For abelian varieties, Matsusaka [56] gave the solution, and for arbitrary groups, it is due to Barsotti [6]. One should also note that Rosenlicht [71] determined the correct field of definition when the ground field contains a dense set of points.

On the other hand, Nakano [66] extended Weil's construction for group varieties to the construction of factor groups and homogeneous spaces, and Weil [90], [91], [92] takes up this question once more in full generality, and gives a self-contained elementary treatment of this topic, giving the construction of transformation spaces and homogeneous spaces over the field k where the birational hypotheses have been made. Weil's method yields abstract varieties. Starting from an idea of Lefschetz, Weil shows how one obtains the projective embedding trivially *a posteriori* for abelian varieties [93] and Chow [18] extends this result to homogeneous spaces. It is the strongest result available at this time.

In order to obtain the desired field of definition, Weil [92] bases himself on a fundamental result of Chow ([14] Th. 3) and reformulates the criteria that can be used to lower the field of definition of a variety. It is those criteria which we have used here. One should compare them with Chatelet's [12] and Lang's [42]. The idea of taking symmetric products is due to Matsusaka [56] and Chow, but in [92] Weil replaces it by a general procedure which can also be used to construct quotient varieties. In the case that he considers, he takes the quotient of a constant field extension.

We take the opportunity of mentioning here the papers of Barsotti [3] and Rosenlicht [72] on the foundation of the theory of algebraic groups, where one can find the fundamental structure theorems. They are not used in this book.

The results of § 2 concerning intersections are due principally to Weil [85]. In [48], Lang points out the advantage of defining explicitly the Pontrjagin products, especially modulo numerical equivalence, in order to handle numerical questions on an abelian variety.

CHAPTER II

General Theorems on Abelian Varieties

An *abelian variety* is a group variety, which, as a variety, is complete. In the classical case, it is not difficult to show that topologically an abelian variety is a complex torus.

We have collected in this chapter various theorems concerning rational maps of varieties into abelian varieties. After § 1, which provides us with some fundamental tools, we take up the study of the Jacobian of a curve. On the one hand, we get the algebraic formulation of Abel's theorem, which describes the group of divisor classes of the curve for linear equivalence by giving to the divisor classes of degree 0 the structure of an abelian variety. On the other hand, we see that the Jacobian, together with a canonical mapping of the curve into it, has the universal mapping property for rational maps of the curve into abelian varieties.

We then extend this second property to varieties of arbitrary dimension, and thus construct what is known as the Albanese variety. In the classical case, this is achieved in the following manner. Assume that V is complete and non-singular. Let $\omega_1, \ldots, \omega_g$ be a basis for the differential forms of first kind on V. (The number g is equal to one half the first Betti number, and is the genus in the case of curves.) If $\mathfrak{a} = \sum n_i P_i$ is a 0-cycle of degree 0 on V, and P is a fixed point of V, then the map into complex g-space given by the vector integral

$$\mathfrak{a} \to \sum n_i \left(\int_P^{P_i} \omega_1, \ldots, \int_P^{P_i} \omega_g \right)$$

is well defined modulo the periods, i.e., vector integrals around topological 1-cycles on V. It can be shown that it induces a homomorphism of the group of 0-cycles of degree 0 onto a complex torus of dimension g, i.e., topologically a product of $2g$ circles. In the case of curves, the kernel consists of those cycles (divisors) that are linearly equivalent to 0; this is Abel's theorem. In higher dimension, one does not know yet a good geometric characterization of the kernel.

The study of the divisor classes is postponed to Chapters III and IV.

As a matter of notation, we shall always use A, B to denote abelian varieties.

§ 1. *Rational maps of varieties into abelian varieties*

THEOREM 1. *An abelian variety A is commutative.*

Proof: Let k be a field of definition for A, and x, y two independent generic points of A over k. Let T be the locus of (x, yxy^{-1}) over k. Geometrically speaking, it is the projection on the second factor $A \times A$ of the graph of the rational map $(x, y) \to (x, yxy^{-1})$, the projection being understood in the usual way. It is the closure in the Zariski topology of the set-theoretic projection. Then T contains the diagonal, and it will suffice to show that $\dim T \leqq \dim A$. Let e be the unit element of A. On $A \times A$ we consider the intersection of T with the variety $e \times A$. This intersection certainly contains (e, e). We contend that it contains nothing else. Let (e, a) be a point in it. Since A is complete, there exists a point (e, b) in $A \times A$ which maps onto (e, a) under the mapping $(x, y) \to (x, yxy^{-1})$. It follows that (e, a) must be equal to (e, e), and that our intersection consists of one point only. Since this point is simple on $A \times A$, the dimension theorem gives

$$0 \geqq \dim T + \dim (e \times A) - 2 \cdot \dim A$$

and consequently $\dim T \leqq \dim A$. This proves our theorem.

We note that a similar argument shows that an abelian subvariety of a group variety is contained in the center.

In view of Theorem 1, the law of composition on an abelian variety will be written additively, and the unit element will be denoted by 0.

THEOREM 2. *Let $f : V \to A$ be a rational map of a variety into an abelian variety. Then f is defined at every simple point of V.*

This theorem will be a consequence of the following stronger result.

LEMMA 1. *Let G be a group variety, and* $f: V \to G$ *a rational map of a variety V into G. Let P be a simple point of V where f is not defined. Then there exists a subvariety W of V, of codimension* 1 *(a divisor), containing P, and such that f is not defined at W.*

Indeed, we know that a rational map of a variety into a complete group variety is defined at every simple subvariety of codimension 1 because the local ring of this subvariety is a valuation ring. It is therefore clear that Lemma 1 implies our theorem.

We shall now prove Lemma 1. Let $F: V \times V \to G$ be the rational map defined by $F(M, N) = f(M)f(N)^{-1}$. I contend that f is defined at P if and only if F is defined at (P, P). Suppose that F is defined at (P, P), and let N be a generic point of V. Then F is *a fortiori* defined at (P, N) and we may write

$$f(P) = F(P, N)f(N).$$

This shows that f is defined at P. The converse is obvious.

If f is defined at P, and hence F at (P, P), then F takes the value e at (P, P). Let G_0 be an affine representative of G on which e has a representative e_0, and let F_0 be the rational map of $V \times V$ into G_0 induced by F. Then the above remarks show that f is defined at P if and only if F_0 is defined at (P, P), and takes the value e_0 at that point.

Let φ_i be the affine coordinate functions of F_0. In order that F_0 (and hence f) be defined at (P, P), it is necessary and sufficient that all the φ_i be defined at (P, P). Hence, if f is not defined at P, one knows from elementary algebraic geometry (IAG—VI_1 Prop. 3) that one of the polar divisors $(\varphi_i)_\infty$ for some φ_i must contain (P, P). Let W be one of the poles of such a function, passing through (P, P). We have $\dim W = 2 \cdot \dim V - 1$. In addition, it is clear that W cannot contain the diagonal. According to the dimension theorem, each component of the intersection $W \cap \Delta$ has dimension $\geqq \dim V - 1$, and F_0 is not defined at such a component. From what we have said above, it follows that f cannot be defined on the projection of such a component on V. This proves Lemma 1, and hence Theorem 2.

The following lemma will be used in Theorem 3.

LEMMA 2. *Let* $f : V \times W \to A$ *be a rational map of a product into an abelian variety. Let* P *be a simple point of* V. *If* f *induces a constant mapping on* $P \times W$, *then there exists a rational map* $f_0 : V \to A$ *such that* $f(M, N) = f_0(M)$. *If* k *is a field of definition for* f, *then* f_0 *is defined over* k.

Proof: We first observe that f is necessarily defined at $P \times W$ according to Theorem 2. We can therefore speak of the induced mapping. We see immediately that our theorem is birational in W, and is geometric. We are going to prove it by induction on the dimension of W.

In the case where W is a curve, we may assume that W is complete, non-singular, because we can replace W if necessary by its normalization over any perfect field. Let Γ be the graph of f, and let T be the projection of Γ on $V \times A$. Let (P, a) be a point of the intersection $T \cap (P \times A)$. Since W is complete, there exists a point (P, Q, a) in Γ projecting on (P, a). Since W is non-singular, the map f is defined at (P, Q), and hence takes on the constant value of f on $P \times W$. This shows that the point a must be equal to this constant, and that our intersection consists of one point. According to the dimension theorem, we have $\dim T \leqq \dim V$, and consequently $\dim T = \dim V$. If (M, N) is a generic point of $V \times W$ over a suitable field k, and $z = f(M, N)$ then (M, z) is a generic point of T, and z must be algebraic over $k(M)$. Since $k(M, N)$ is a regular extension of $k(M)$, it follows that z is rational over $k(M)$, and hence that z can be written $f_0(M)$ for some rational map f_0. This concludes the proof in case W is a curve.

We are now going to prove the general case by considering an algebraic system on W, for which both the varieties of the system and the parameter variety will be of lower dimension. Algebraically, and field theoretically, we express this as follows.

Let k be an algebraically closed field of definition for f, and let (M, N) be a generic point of $V \times W$ over k. Let $k(t)$ be an intermediate field between k and $k(N)$ such that $k(t)$ has dimension 1 over k, and $k(N)$ is a regular extension of $k(t)$. (Geometrically,

$k(t)$ is the field of parameters t, and the variety U which is the locus of N over $k(t)$ is the generic element of a pencil on W.) The existence of such a field is easy to prove, and we shall recall the argument later for the convenience of the reader. We have $\dim U = \dim W - 1$, and the rational map f induces a rational map f' on $V \times U$, such that $f'(M, N) = f(M, N)$. Naturally, f' is defined over $k(t)$. By the induction hypothesis, there exists a rational map $f_0 : V \times T \to A$, where T is the locus of t over k, such that

$$f_0(M, t) = f(M, N).$$

We have thus reduced the proof to the case of curves, which has already been dealt with.

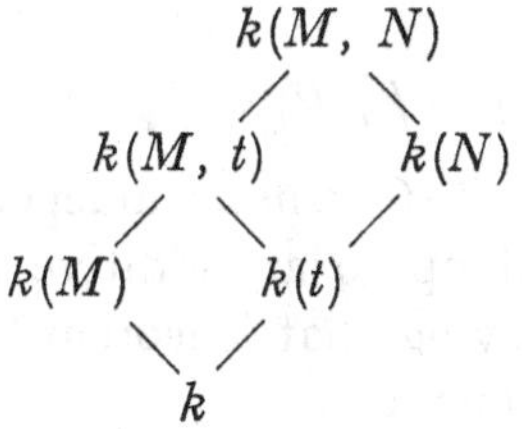

There are many ways of proving the existence of the desired field $k(t)$ used above. Here is one of them.

LEMMA 3. *Let k be a perfect field, and K a finitely generated regular extension. Let E be an intermediate field between k and K having transcendence degree 1 over k, and algebraically closed in K. Then K is a regular extension of E.*

Proof: We must show that K is separable over E, or in other words that K is linearly disjoint from $E^{1/p}$ over E. Since k is perfect, $E^{1/p}$ can be generated by one element over E. In fact, one sees immediately that if x is a separating transcendental element of E over k, then $E^{1/p} = E(x^{1/p})$. The irreducible equation for $x^{1/p}$ over E must then remain irreducible over K since E is algebraically closed in K. This proves the linear disjointness.

THEOREM 3. *Let $f : V \times W \to A$ be a rational map of a product into an abelian variety. Then there exist two rational maps $f_1 : V \to A$*

and $f_2 : W \to A$ *such that for* (P, Q) *generic on* $V \times W$ *we have* $f(P, Q) = f_1(P) + f_2(Q)$, *and* f_1, f_2 *are uniquely determined by this property, up to an additive constant. Furthermore, if* f *is defined over* k, *and if* V *has a simple point rational over* k, *then we can take* f_1 *and* f_2 *defined over* k.

Proof: Let P' be a simple point of V rational over k. According to Theorem 2, f induces a rational map of $P' \times W$ into A. Put $f_1(Q) = f(P', Q)$ and apply Lemma 2 to the map $f(P, Q) - f(P', Q)$ which is constant on $P' \times W$. There exists a rational map $f_2 : W \to A$ defined over k such that $f(P, Q) = f_1(P) + f_2(Q)$.

As to the uniqueness, suppose that we can write

$$f(P, Q) = f_1(P) + f_2(Q) = f'_1(P) + f'_2(Q).$$

We get

$$f_1(P) - f'_1(P) = f'_2(Q) - f_2(Q).$$

Since we can select P, Q generic independent, this shows that f_1, f_2 are determined up to an additive constant.

It goes without saying that Theorem 3 can be generalized to a finite number of factors.

We remark parenthetically that Weil's theorems on homogeneous spaces [91] show that there exists a principal homogeneous space H of A, defined over k, and two rational maps $\varphi_1 : V \to H$ and $\varphi_2 : W \to H$ also defined over k such that, if $\varphi : H \times H \to A$ is the canonical map, we have $f = \varphi(\varphi_1, \varphi_2)$.

One of the most remarkable applications of Theorem 3 is the following theorem.

THEOREM 4. *Every rational map of a group variety* G *into an abelian variety* A *is a homomorphism, up to a translation. More precisely, if* $f : G \to A$ *is such a rational map, defined over* k, *and if we put* $a = f(e)$, *then there is a homomorphism* $f_0 : G \to A$ *defined over* k *such that* $f(x) = f_0(x) + a$.

Proof: Consider the rational map $F : G \times G \to A$ defined by the formula $F(x, y) = f(xy)$. After a suitable translation, we may assume that $f(e) = 0$. The rational map of $G \times G$ into A given by the formula $F(x, y) - F(e, y)$ is constant on $e \times G$. Using either Theorem 2 or Theorem 3, we see that there exists a rational map $g : G \to A$ defined over k such that

$$f(xy) = f(y) + g(x).$$

Putting $y = e$, we see that $g = f$, thereby proving our theorem.

We shall say that a variety V is *pure* (resp. *semi-pure*) if its function field over some field of definition is a purely transcendental extension (resp. is contained in a purely transcendental extension) of the constant field.

COROLLARY. *Every rational map of a pure, or semi-pure, variety into an abelian variety is constant.*

Proof: Since the corollary is birational with respect to the variety, we may assume that it is a product of straight lines. Using Theorem 3, we may therefore assume that V is of dimension 1, and is either the affine line viewed as a group variety under addition, or the multiplicative group. According to Theorem 4, there exist two constants a, b such that $f(x + y) = f(x) + f(y) + a$ and $f(xy) = f(x) + f(y) + b$, where x, y are two independent generic points of V, and $x + y$, xy denote the addition on the affine line and the product, respectively. This is obviously absurd unless f is constant.

Let A, B be two abelian varieties. The homomorphisms of A into B form an additive group, which will be denoted by $H(A, B)$. In case $A = B$, we denote $H(A, A)$ by End (A), and its elements will be called *endomorphisms*.

Let A, B, C be three abelian varieties. Let $\alpha : A \to B$ and $\beta : B \to C$ be two homomorphisms. Then their composition $\beta\alpha : A \to C$ as rational maps is obviously a homomorphism of A into C. In particular, if $A = B = C$, $\beta\alpha$ is an endomorphism of A. This product obviously determines a ring structure on End (A). Its identity will be denoted by δ_A, or simply by δ if the reference to A is clear. We shall show in Chapter III that $H(A, B)$ and End (A) are without torsion, i.e., that if $\alpha : A \to B$ is a homomorphism, and $\alpha \neq 0$, then for any integer $n \neq 0$ we have $n \cdot \alpha \neq 0$.

Let α be surjective. As usual we denote by $\nu(\alpha)$ the degree of α

if it is finite and 0 otherwise. We can say that if $\nu(\alpha) \neq 0$ then $\nu(n\alpha) \neq 0$, or $\nu(n\delta) \neq 0$ whenever $n \neq 0$.

The tensor products of $H(A, B)$ and End (A) with the rational numbers $\mathbf{Q}$ will be denoted by $H_{\mathbf{Q}}(A, B)$ and $\text{End}_{\mathbf{Q}}(A)$, and $\text{End}_{\mathbf{Q}}(A)$ will be called the *algebra of endomorphisms* of A.

In the next theorem (called *Chow's theorem*) we shall use the hypothesis $\nu(n\delta) \neq 0$ in order to deal with a technical point due to characteristic p. No logical confusion will occur, since that part of the theorem which depends on this hypothesis will not be used until Chapter V, and the proof that $\nu(n\delta) \neq 0$ will be given in Chapter IV.

Chow's theorem expresses in algebraic terms the classical fact that an abelian variety does not contain an algebraic system of abelian subvarieties.

THEOREM 5. *Let A be an abelian variety defined over a field k, and let B be an abelian subvariety of A defined over a primary extension K of k. Then B is defined over k.*

Proof: We recall that an extension K of k is said to be primary if the algebraic closure of k in K is purely inseparable. If k' is a suitable purely inseparable extension of k, then Kk' is regular over k'. Hence to prove our theorem, we may deal separately with the two cases where K is regular over k, and where K is purely inseparable over k.

Suppose first that K is regular over k. In this part of the proof, we make no use of the fact that $\nu(n\delta) \neq 0$. We may assume without loss of generality that K is finitely generated over k. Write $K = k(t)$ where t is the generic point of some parameter variety T, and suppose that there exists an abelian subvariety $B = B_t$ of A defined over $k(t)$. We must show that B_t is defined over k. Let t' be an independent generic specialization of t over k, and take the sum $B_t + B_{t'}$, which is defined over $k(t, t')$. It is an abelian subvariety of A, which is equal to B if and only if B is defined over k. Indeed, we have $B_t \subset B_t + B_{t'}$. If $B_t = B_t + B_{t'}$ then $B_t = B_{t'}$, and B_t is defined over $k(t) \cap k(t') = k$. The converse is obvious.

Let now u, u' be two independent generic points of B_t, $B_{t'}$ over $k(t, t')$, and let $\Gamma = \mathrm{loc}_k\,(u + u', t, t')$. Then Γ is a subvariety of $A \times T \times T$, and it will suffice to show that $\dim \Gamma \leqq \dim B + 2 \cdot \dim T$, because this will imply that the dimension of $u + u'$ over $k(t, t')$ is $\leqq \dim B$, and hence that $u + u' \in B$.

We consider the intersection $\Gamma \cap (A \times t \times t)$. Let (v, t, t) be a point in this intersection. We are going to prove that $v \in B$. The specialization $\sigma : (t, t') \to (t, t)$ induces an isomorphism on each one of the fields $k(t)$ and $k(t')$, and we get $B_t{}^{\sigma} = B_t$, $B_{t'}{}^{\sigma} = B_t$. Since B is complete, we can find two points u_1, $u_2 \in B$ such that $v = u_1 + u_2$, whence $v \in B$. The dimension of each component of $\Gamma \cap (A \times t \times t)$ taken on $A \times T \times T$ is therefore $\leqq \dim B$. The dimension theorem gives us $\dim \Gamma \leqq \dim B + 2 \cdot \dim T$, and this shows that the dimension of $u + u'$ over $k(t, t')$ is at most equal to the dimension of B. This proves the theorem in the case where K is regular over k.

We now assume that K is purely inseparable over k and of degree p^m, and in this part of the proof we make use of the hypothesis that the kernel of $p^e\delta$ is finite. Let u be a point of A. The Frobenius automorphism $\xi \to \xi^{p^m}$ of the universal domain transforms A into an abelian variety denoted by $A^{(p^m)}$ and u into $u^{(p^m)}$. If u is a generic point of B over K, then $u^{(p^m)}$ is a generic point of $B^{(p^m)}$ over $K^{(p^m)} \subset k$. This shows that $B^{(p^m)}$ is defined over k, and hence that $k(u^{(p^m)})$ is a regular extension of k. One sees immediately that $K(u)$ is a purely inseparable extension of $k(u^{(p^m)})$. Let p^e be its degree. Then the cycle $p^e \cdot (u)$ is rational over $k(u^{(p^m)})$ and according to the fundamental theorem on symmetric functions (Ch. I, § 1) it follows that the point $p^e u$ is rational over that field. Hence $k(p^e\, u)$ is a regular extension of k, since it is contained in $k(u^{(p^e)})$. Furthermore $p^e u$ is a point of B, and is even a generic point of B over k because the kernel of $p^e\delta$ is finite. This shows that B is defined over k, and concludes our proof.

We shall end this section with another application of the hypothesis $\nu(n\delta) \neq 0$, namely Poincaré's theorem of complete reducibility. As for the preceding theorem, we shall not use this result until Chapter V.

THEOREM 6. *Let A be an abelian variety, B an abelian subvariety of A. Then there exists an abelian subvariety C of A such that $A = B + C$, and $B \cap C$ is a finite group. If A, B are defined over a field k, then we can take C also defined over k.*

Proof: Let $\lambda : A \to A/B$ be the canonical homomorphism of A onto the factor group, which we denote by H. The idea of the proof is to define a cross section, or an approximate cross section of H into A. Let u be a generic point of A over k, and put $\lambda(u) = v$. Let W be the variety $\lambda^{-1}(v)$. It is the locus of u over $k(v)$. Then $W = B_u$ is the translation $B + u$ of B by u. It is a homogeneous space for B. If W has a rational point over $k(v)$, then we can define the above mentioned section, and the theorem would be proved. In general, we are going to use an approximate section. More precisely, let P be a point of W which is rational over a separable algebraic extension of $k(v)$, and let P_i be its conjugates over $k(v)$. We can write $P_i = Q_i + u$ with $Q_i \in B$. Take the sum on A. We get

$$\sum P_i = \sum Q_i + d \cdot u.$$

This shows that $\sum Q_i$ is a point of B, rational over $k(u)$. We denote it by y. There exists a rational map $\varphi : A \to B$ defined over k, such that $\varphi(u) = y$. Let x be a generic point of B over $k(u)$. Then $u + x$ is a generic point of W over $k(u)$, and we have an isomorphism of $k(u)$ over $k(v)$ mapping u on $u + x$. Geometrically, this means that we have made B operate generically on the homogeneous space W. Since the P_i are algebraic over $k(v)$, there exists an extension of this isomorphism which leaves the P_i fixed. This extension maps

$$y = \sum Q_i = \sum P_i - d \cdot u$$

on

$$\sum P_i - d \cdot u - d \cdot x$$

and we have therefore $\varphi(u + x) = \varphi(u) - d \cdot x$. We know that φ is everywhere defined (Theorem 2). Putting $u = 0$, we obtain

$\varphi(x) = - d \cdot x$, which shows that the restriction of φ to B is surjective, and that its kernel in A meets B only in a finite number of points. Let C be the connected component of this kernel. Then C is defined *a priori* over a purely inseparable extension of k, because C is k-closed. Theorem 5 shows that C is defined over k. We have $\dim B + \dim C = \dim A$, and $C \cap B$ is finite. The kernel of the homomorphism $B \times C \to B + C$ is finite, and we thus see that $A = B + C$. This proves the theorem.

REMARK. More generally, we may suppose that B is an abelian subvariety of any group variety G. Essentially the same proof shows that there exists a group subvariety G_1 of G such that $G = G_1 B$ and $G_1 \cap B$ is finite. We note that in this case, G_1 may be defined only over a purely inseparable extension of k. Cf. Rosenlicht [72].

Let A, B be two abelian varieties. We shall say that A is *isogenous* to B if there exists a surjective homomorphism $\lambda : A \to B$ whose kernel is finite. The relation of isogenity is obviously transitive and reflexive. To show that it is symmetric, let n be the degree of λ, and suppose that λ is defined over k. Let v be a generic point of B over k, and let $\sum (u_i)$ be the cycle $\lambda^{-1}(v)$. If λ is not separable, then each point in this cycle appears with multiplicity > 1. Put $w = \sum \lambda u_i$, the sum being taken on A. Then the fundamental theorem on symmetric functions shows that w is rational over $k(v)$, and there exists a homomorphism $\beta : B \to A$ such that $\beta(v) = w$. Since $\lambda u_i = v$ for each u_i, we find $\beta\lambda = n\delta$. Under the hypothesis that $\nu(n\delta) \neq 0$, we conclude that the relation of isogenity is an equivalence relation. For every isogeny $\lambda : A \to B$ we can find an isogeny $\beta : B \to A$ such that $\beta\lambda = \nu(\lambda) \cdot \delta$.

In the tensor product $H_{\mathbf{Q}}(B, A)$, the element $1/\nu(\lambda) \cdot \beta$ is called the *inverse* of λ and sometimes denoted by λ^{-1}. If $\lambda \in H_{\mathbf{Q}}(A, B)$, and $\nu(\lambda) \neq 0$, then λ has an inverse in $H_{\mathbf{Q}}(B, A)$.

We shall say that an abelian variety A is *simple* if A and 0 are the only abelian subvarieties of A. Theorem 6 can be interpreted by saying that for each abelian subvariety B of A we can find an abelian subvariety C of A such that A is isogenous to the product $B \times C$. This reducibility property together with an obvious descending chain condition leads to the following result.

COROLLARY. *Every abelian variety is isogenous to a product of simple abelian varieties, uniquely determined up to isogenies.*

We are thus faced with a situation which is formally analogous to that of Wedderburn's theorem. The ring of endomorphisms of a simple abelian variety has no divisors of 0, and the algebra of endomorphisms $\mathrm{End}_{\mathbf{Q}}(A)$ of a simple abelian variety is therefore a field (not necessarily commutative).

If A is any abelian variety, and $\lambda : A \to B$ is an isogeny, then $\mathrm{End}_{\mathbf{Q}}(A)$ and $\mathrm{End}_{\mathbf{Q}}(B)$ are isomorphic, under the mapping

$$\alpha \to \lambda\alpha\lambda^{-1}$$

for $\alpha \in \mathrm{End}_{\mathbf{Q}}(A)$. We know that A is isogenous to a product

$$(A_1 \times \ldots \times A_1) \times \ldots \times (A_m \times \ldots \times A_m)$$

where the A_i are simple abelian varieties, mutually non-isogenous. We then obtain the following theorem.

THEOREM 7. *Let A be an abelian variety, isogenous to a product as above, and suppose that A_i occurs n_i times, $i = 1, \ldots, m$. Let H_i be the complete ring of matrices of degree n_i over the field* $\mathrm{End}_{\mathbf{Q}}(A_i)$. *Then* $\mathrm{End}_{\mathbf{Q}}(A)$ *is isomorphic under the obvious natural mapping to the direct product of the rings H_i.*

It will be proved in Chapter VII that the additive group of endomorphisms of an abelian variety is of finite type, and hence that the algebra of endomorphisms is semi-simple.

§ 2. *The Jacobian variety of a curve*

Let C be a complete, non-singular curve of genus $g > 0$, defined over a field k. All the fields in the rest of this section are assumed to contain k. We are going to construct the Jacobian of C, and to begin with, we shall use the Riemann-Roch theorem to define a normal law of composition on the field of symmetric functions of the product of C with itself g times.

LEMMA 4. *Let U be a variety defined over k, and let $P_1, \ldots, P_n$ be independent generic points of U over k. Then the smallest field* *of rationality of the cycle* $\sum P_i$ *is the field* $k(P_1, \ldots, P_n)_s$ *of symmetric functions, i.e., the subfield of* $k(P_1, \ldots, P_n)$ *which is left fixed by the group of permutations of the* P_i. *The cycle* $\sum P_i$ *is a prime rational cycle over that field.*

Proof: Put $K = k(P_1, \ldots, P_n)_s$. Every automorphism of the universal domain leaving K fixed leaves $\sum P_i$ invariant, and hence $\sum P_i$ is rational over a purely inseparable extension of K. Since $k(P_1, \ldots, P_n)$ is a separable extension of K (and even a Galois extension, according to Galois theory), it follows that our cycle is rational over K. Let L be the smallest field of rationality of $\sum P_i$. Then $L \subset K$. Every automorphism of the universal domain leaving L fixed leaves $\sum P_i$ invariant, and hence K fixed. Hence K is purely inseparable over L. Since each P_i has coefficient 1, the extension $L(P_1, \ldots, P_n)$ of L is separable, and hence K is separable over L, whence $K = L$. In addition, $\sum P_i$ is prime rational over K, because the automorphisms of $k(P_1, \ldots, P_n)$ leaving K fixed permute the P_i transitively.

LEMMA 5. *Let* $\mathfrak{a}$ *be a divisor of degree* 0 *on* C, *and let* $P_1, \ldots, P_g$ *be* g *independent generic points of* C *over a field* K *over which* $\mathfrak{a}$ *is rational. Then there exists one and only one positive divisor on* C *linearly equivalent to* $\mathfrak{a} + \sum P_i$, *and this divisor is of type* $\sum Q_i$ *where the* Q_i *are independent generic points of* C *over* K. *Furthermore, we have*

$$K(P_1, \ldots, P_g)_s = K(Q_1, \ldots, Q_g)_s.$$

Proof: We begin by some remarks concerning the Riemann-Roch theorem. Let $\mathfrak{b}$ be a divisor on C, and denote as usual by $l(\mathfrak{b})$ the dimension of the vector space $L(\mathfrak{b})$ of functions f on C such that $(f) \geqq -\mathfrak{b}$. The Riemann-Roch theorem asserts that

$$l(\mathfrak{b}) = \deg \mathfrak{b} + 1 - g + \delta(\mathfrak{b}),$$

and $\delta(\mathfrak{b}) = l(\mathfrak{c} - \mathfrak{b})$ for any divisor $\mathfrak{c}$ of the canonical class. We may select $\mathfrak{c}$ rational over k. If $\delta(\mathfrak{b}) \neq 0$, if $\mathfrak{b}$ is rational over a field E, and if P is a generic point of C over E, then $\delta(\mathfrak{b} + P) = \delta(\mathfrak{b}) - 1$. This is easily seen, for instance, as follows. It is trivial that if $\delta(\mathfrak{b} + P) \neq \delta(\mathfrak{b}) - 1$, then $\delta(\mathfrak{b} + P) = \delta(\mathfrak{b})$. This implies that all the functions of $L(\mathfrak{c} - \mathfrak{b})$ have a zero at P. Since $\mathfrak{c}$ can be selected rational over E, this space of functions has a basis defined over E, and we thus get a contradiction.

Now put $\mathfrak{p} = \sum P_i$. From the hypothesis $\deg \mathfrak{a} = 0$, we see immediately that $l(\mathfrak{a}) = 0$ or 1. Using the above remarks, and adding one generic point after another to $\mathfrak{a}$, we find $l(\mathfrak{a} + \mathfrak{p}) = 1$. Indeed, if $l(\mathfrak{a}) = 0$ then $\delta(\mathfrak{a}) = g - 1$ and hence $\delta(\mathfrak{a} + \mathfrak{p}) = 0$, and if $l(\mathfrak{a}) = 1$, then $\delta(\mathfrak{a}) = g$, and hence we still have $\delta(\mathfrak{a} + \mathfrak{p}) = 0$. Since the divisor $\mathfrak{a} + \mathfrak{p}$ is rational over $K(P_1, \ldots, P_g)_s$, the unique positive divisor $\mathfrak{q}$ in the complete linear system of $\mathfrak{a} + \mathfrak{p}$ is rational over that field. Conversely, we see that $\mathfrak{p}$ is the only positive divisor in the linear equivalence class of $\mathfrak{q} - \mathfrak{a}$, and hence that $\mathfrak{p}$ and $\mathfrak{q}$ have the same smallest field of rationality containing K. Since the transcendence degree of this field over K is equal to g, and since $\mathfrak{q}$ has degree g, we conclude that $\mathfrak{q}$ is of the desired type, thereby proving our lemma.

To go further, we shall assume that C has a positive divisor $\mathfrak{o}$ of degree g, rational over k. This condition is satisfied if C has a rational point over k. It will allow us to construct the Jacobian over k.

We are now going to define a normal law of composition on the field $k(P_1, \ldots, P_g)_s$. Let V be any model of this field, defined over k. Let C^g denote the (ordinary) product of C with itself g times. We have a rational map $F : C^g \to V$ of degree $g!$. If u is a generic point of V over k, then the cycle $F^{-1}(u)$ is uniquely determined as a cycle of type $\sum P_i$, where the P_i are g independent generic points of C over k, and the smallest field of rationality of $\sum P_i$ is precisely $k(u)$. We may therefore write $u = F(\mathfrak{p})$ with $\mathfrak{p} = \sum P_i$.

Let P_i, Q_j be $2g$ independent generic points of C over k, and put $\mathfrak{p} = \sum P_i$, $\mathfrak{q} = \sum Q_j$. If we take $\mathfrak{a} = \mathfrak{q} - \mathfrak{o}$, then Lemma 5 shows that

$$l(\mathfrak{p} + \mathfrak{q} - \mathfrak{o}) = 1$$

and that

$$\mathfrak{p} + \mathfrak{q} - \mathfrak{o} \sim \mathfrak{m}$$

for some positive divisor $\mathfrak{m} = \sum M_i$, uniquely determined by $\mathfrak{p}$ and $\mathfrak{q}$, consisting of g independent generic points of C over k. Put $u = F(\mathfrak{p})$, $v = F(\mathfrak{q})$, and $w = F(\mathfrak{m})$. Using the additive notation, the law of composition on V is then defined by $u + v = w$. This relation gives us a rational map of $V \times V$ into V, because $\mathfrak{m}$ is rational over the field $k(u, v)$ which is a field of rationality for $\mathfrak{p} + \mathfrak{q} - \mathfrak{o}$, and hence w is rational over $k(u, v)$. The rational map is obviously generically surjective. Conversely, the uniqueness of the positive divisor in the class of $\mathfrak{m} + \mathfrak{o} - \mathfrak{q}$ shows that $\mathfrak{p}$ (and hence u) is rational over $k(v, w)$. This symmetry shows that we have

$$k(u, v) = k(u, w) = k(v, w).$$

Finally, if $\mathfrak{p}_1, \mathfrak{p}_2, \mathfrak{p}_3$ are three generic independent cycles of degree g, of the preceding type, then looking at the linear equivalence class of

$$\mathfrak{p}_1 + \mathfrak{p}_2 + \mathfrak{p}_3 - 2 \cdot \mathfrak{o}$$

we see immediately that our law of composition is associative. The commutativity is obvious.

From Chapter I, § 1 we conclude that we can select our model V as a commutative group variety which will be denoted by J, and which is called the *Jacobian* of C. It is defined over k. To show that it is an abelian variety, we use the following proposition.

PROPOSITION 1. *Let C be a complete non-singular curve. Let $\mathfrak{a}$ be a divisor on C which is linearly equivalent to* 0. *Then every specialization of $\mathfrak{a}$ is also linearly equivalent to* 0.

Proof: It is well known and easy to prove that if $\mathfrak{a}'$ is a specialization of $\mathfrak{a}$ then there exists a curve T, a divisor X on $T \times C$ and a field K having the following property. T is defined over K, X is rational over K, and there is a generic point t of T and a simple point t' of T such that $\mathfrak{a} = X(t) = \mathrm{pr}_2[X \cdot (t \times C)]$ and $\mathfrak{a}' = X(t') = \mathrm{pr}_2[X \cdot (t' \times C)]$ (see for instance Matsusaka [55]). We can then use first Theorem 6 in the Appendix (with $Y = 0$), to conclude that X is degenerate. Using then the Corollary of Proposition 5 and Proposition 6 of the Appendix, we get what we want.

We can now prove that J is an abelian variety. Let u, v be two independent generic points of J over k. Let φ be a place of $k(u)$ which is $\bar{k}$-valued. We can extend it to a place of $k(u, v)$ which leaves $k(v)$ fixed, because $k(u)$ and $k(v)$ are linearly disjoint over k. We must show that φ induces a point on J, i.e., does not send u to infinity. Write $u = F(\mathfrak{p})$ and $v = F(\mathfrak{q})$. We have

$$\mathfrak{p} + \mathfrak{q} - \mathfrak{o} \sim \mathfrak{m}.$$

Extend φ to a place of $k(P_1, \ldots, P_g, Q_1, \ldots, Q_g)$, and put $P'_i = \varphi(P_i)$, $M'_i = \varphi(M_i)$. The points Q_i are permuted by φ, since φ leaves $k(v)$ fixed. The cycles

$$\mathfrak{p}' = \sum P'_i, \quad \mathfrak{q}' = \mathfrak{q}, \quad \mathfrak{m}' = \sum M'_i$$

are specializations of $\mathfrak{p}$, $\mathfrak{q}$, and $\mathfrak{m}$, respectively, over k. According to Proposition 1, we have

$$\mathfrak{p}' + \mathfrak{q} - \mathfrak{o} \sim \mathfrak{m}'.$$

In addition, $\mathfrak{p}'$ is rational over $\bar{k}$, and $\mathfrak{q}$ is generic over $\bar{k}$. Hence $\mathfrak{m}'$ is generic over $\bar{k}$ by Lemma 5. If we put $w = F(\mathfrak{m})$ and $w' = F(\mathfrak{m}')$, then $w' = \varphi(w)$, and we see that our place induces a specialization $(v, w) \to (v', w')$ over k, where w' is a generic point of J over k. Since we have $u = v - w$ on J, and since $\varphi(v)$ and $\varphi(w)$ are points of J, it follows that $\varphi(u)$ is a point of J, and hence that J is complete: it is an abelian variety.

REMARK. The construction preceding Proposition 1 is essentially formal, and with suitable modifications has been applied by Rosenlicht to construct the generalized Jacobians [71]. It is through Proposition 1 that we have used the hypothesis that C is complete and non-singular, to show that the Jacobian is also complete.

We have obtained above a rational map $F : C^g \to J$. By Theorem 2 of § 1 we know that this rational map is everywhere defined, because the product C^g is non-singular. By Theorem 3 of § 1, we can decompose F into a sum of mappings of each factor. By symmetry, these mappings are all equal. In other words, there exists a rational map $f : C \to J$ and a constant $c \in J$ such that

$$F(P_1, \ldots, P_g) = \sum f(P_i) + c.$$

The mapping f is uniquely determined up to a translation, and will be called the *canonical mapping* of C into J. If C has a rational point over k, then f is defined over k, and c is rational over k.

We summarize the preceding construction as follows.

THEOREM 8. *Let C be a complete non-singular curve of genus $g > 0$. Then there exists an abelian variety J of dimension g, and a rational map $f : C \to J$ such that if K is a field of definition for f, and $P_1, \ldots, P_g$ are independent generic points of C over K, then the point*

$$u = \sum f(P_i)$$

is a generic point of J over K and $K(u) = K(P_1, \ldots, P_g)_s$. Conversely, every generic point of J can be expressed in this manner, and the P_i are uniquely determined, up to a permutation. Finally, if C has a rational point over k, we can select f and J to be defined over k.

We note that the conditions of Theorem 8 characterize J up to a birational isomorphism and f up to a translation. Indeed, if f', J' satisfy these conditions, one sees immediately that J' and J are birationally equivalent, and hence birationally isomorphic according to Theorem 4. We can therefore put $J = J'$. As we have already remarked, the map $F : C^g \to J$ defined by $F(P_1, \ldots, P_g) = \sum f(P_i)$ and the map F' defined by means of f' in a similar manner determine f and f' uniquely, up to an additive constant, by Theorem 3 and the fact that F, F' are symmetric on C^g.

The next theorem gives us the universal mapping property of (J, f) with respect to rational maps of C into abelian varieties.

THEOREM 9. *Let $f : C \to J$ be a canonical map of C into its Jacobian. Let $h : C \to A$ be a rational map of C into an abelian variety. Then there exists one and only one homomorphism $\alpha : J \to A$ such that $h = \alpha f + a$, with a suitable constant a. If f, h are defined over k, and if C has a rational point over k, then α is defined over k, and a is rational over k.*

Proof: Let P' be the rational point of C over k. Then $h(P')$

is a point of A, rational over k, and after subtracting $h(P')$ from h, we see that we may assume that $h(P') = 0$. Similarly, we may assume that $f(P') = 0$. Let $P_1, \ldots, P_g$ be independent generic points of C over k. Suppose that we can write $h = \beta f + b$, with a homomorphism β of J into A, and a constant b of A. Put $u = f(P_1) + \ldots + f(P_g)$. It is a generic point of J over k, and after an obvious subtraction, we see that $\alpha u - \beta u$ is constant, and hence that $\alpha = \beta$. This proves the uniqueness of the homomorphism induced by h.

To prove the existence, let $H : C^g \to A$ be the sum of h with itself g times, i.e., $H(P_1, \ldots, P_g) = h(P_1) + \ldots + h(P_g)$. It is symmetric in $P_1, \ldots, P_g$, and the point $v = H(P_1, \ldots, P_g)$ is rational over $k(u)$. Hence there exists a rational map $\alpha : J \to A$ such that $\alpha u = v$. Taking into account Theorem 4, and recalling that $f(P') = h(P') = 0$, we see that α is a homomorphism. This proves our theorem.

Let $f : V \to A$ be a rational map of a variety into an abelian variety. Then f induces a homomorphism of the group of cycles on V into A as follows. We denote by $Z_r(V)$ the group of cycles of dimension r on V. Let $\mathfrak{a} = \sum n_i(x_i)$ be an element of $Z_0(V)$. We put $f(\mathfrak{a}) = \sum n_i(f(x_i))$. It is an element of $Z_0(A)$. Taking the sum on A, we get a point $S(f(\mathfrak{a})) = S_f(\mathfrak{a})$. We shall consider especially cycles of dimension 0 in this section, and we write simply $Z(V)$. By $Z(V, 0)$, we denote the cycles of degree 0. The restriction of S_f to $Z(V, 0)$ induces a homomorphism of $Z(V, 0)$ into A, and the next theorem gives us the kernel in the case of the canonical map of a curve into its Jacobian. It is known as *Abel's theorem.*

THEOREM 10. *Let $f: C \to J$ be a canonical map of C into its Jacobian J, and let S_f be the homomorphism of $Z(C, 0)$ into J. Then the kernel of S_f is equal to the group of divisors on C which are linearly equivalent to 0.*

Proof: Let $\mathfrak{a}$ be a divisor on C which is linearly equivalent to 0. We have $\mathfrak{a} = (\varphi)$ for some function φ of C into the projective straight line. Let K be a field of definition for f and φ, and let

P be a generic point of C over K. Put $t = \varphi(P)$. The divisor $\mathfrak{a}_t = \varphi^{-1}(t)$ is rational over $K(t)$, which is a purely transcendental extension of K. The point $S_f(\mathfrak{a}_t)$ of J is rational over $K(t)$ by the main theorem on symmetric functions, and the map $t \to S_f(\mathfrak{a}_t)$ gives rise to a rational map of a pure variety into J. It is constant according to the corollary of Theorem 4. Since the divisor $\mathfrak{a}$ is equal to the difference $\mathfrak{a}_0 - \mathfrak{a}_\infty$, these divisors being the specializations of $\mathfrak{a}_t$ over $t \to 0$ and $t \to \infty$, respectively, we see that $S_f(\mathfrak{a}) = 0$.

Conversely, let $\mathfrak{a}$ be a divisor on C of degree 0, rational over K, and such that $S_f(\mathfrak{a}) = 0$. Let $P_1, \ldots, P_g$ be independent generic points of C over K, and put $\mathfrak{p} = \sum P_i$. According to Lemma 5, we have $\mathfrak{a} + \mathfrak{p} \sim \mathfrak{q}$, with $\mathfrak{q} = \sum Q_i$, the Q_i being generic independent over K. By hypothesis, and the first part of our theorem, we know that $S_f(\mathfrak{p}) = S_f(\mathfrak{q})$. But we know that a generic point $u = \sum f(P_i)$ of J determines the cycle $\sum P_i$. Hence we must have $\mathfrak{p} = \mathfrak{q}$. This proves the theorem.

One may ask the question whether the set-theoretic image $f(C)$ of C into its Jacobian is birationally, or even biholomorphically equivalent to C. This is indeed the case, and we are going to prove this using intersection theory. It will be the only quantitative result of this chapter.

The sum $f(C) \oplus f(C) \oplus \ldots \oplus f(C)$ taken r times $(1 \leqq r \leqq g)$ will be denoted by W_r. We are now going to determine the Pontrjagin products $f(C) * f(C) * \ldots * f(C)$ (using the notation of Chapter I, § 2)

Proposition 2. *Let $f: C \to J$ be a canonical map of C into its Jacobian, and assume that it is defined over k. Let $P_1, \ldots, P_r$ be independent generic points of C over k, with $1 \leqq r \leqq g$. Put*

$$x = \sum_{i=1}^{r} f(P_i).$$

Then $k(x) = k(P_1, \ldots, P_r)_s$, and

$$[k(P_1, \ldots, P_r) : k(x)] = r!$$

In particular, the map of C into its Jacobian is birational.

Proof: Since the divisor $\sum_{i=1}^{r} P_i$ is rational over $k(P_1, \ldots, P_r)_s$, it follows that x is also rational over this field, and $k(x)$ is therefore contained in this field. According to Lemma 1, it will suffice to prove that $k(P_1, \ldots, P_r)$ is an algebraic extension of $k(x)$ of degree $r!$ because we have the inclusion

$$k(x) \subset k(P_1, \ldots, P_r)_s \subset k(P_1, \ldots, P_r).$$

Let $P_{r+1}, \ldots, P_g$ be generic points of C independent of $P_1, \ldots, P_r$. Put $K = k(P_{r+1}, \ldots, P_g)$ and $y = \sum_{j=r+1}^{g} f(P_j)$. Then $K(x) = K(x + y)$: it is the compositum of $k(x)$ and K.

$$\begin{array}{ccccccc}
 & & k(P_1, \ldots, P_g) & & & & \\
 & / & & \backslash & & & \\
k(P_1, \ldots, P_r) & & & & K(x) = K(x+y) & & \\
 & \backslash & & / & & \backslash & \\
 & & k(x) & & & & K = k(P_{r+1}, \ldots, P_g) \\
 & & & \backslash & & / & \\
 & & & & k & &
\end{array}$$

Furthermore, $K(x)$ and $k(P_1, \ldots, P_r)$ are linearly disjoint over $k(x)$ because K and $k(P_1, \ldots, P_r)$ are linearly disjoint over k. It will therefore suffice to show that $k(P_1, \ldots, P_g)$ is an extension of degree $r!$ over $K(x)$. We see immediately that $K(x + y)$ is contained in the subfield of $k(P_1, \ldots, P_g)$ left invariant by the automorphisms which permute $P_1, \ldots, P_r$ and leave $P_{r+1}, \ldots, P_g$ fixed. It is equal to this field, because on the one hand $k(P_1, \ldots, P_g)$ is separable algebraic over $k(x + y)$ and hence over $K(x + y)$, and on the other hand every isomorphism of $k(P_1, \ldots, P_g)$ which leaves $K(x + y)$ invariant leaves $P_{r+1}, \ldots, P_g$ invariant. We can now use Lemma 1. This concludes the proof.

If we take the sum of $f(C)$ on J $g - 1$ times, we obtain a subvariety of J of codimension 1, i.e., a prime divisor, which we shall denote by Θ. It is well defined up to a translation by C and its canonical map into J. We are now going to determine the intersection $f(C) \cdot \Theta^{-}_{u}$ as a special case of the Pontrjagin products given in Chapter I, § 2. We take u generic over k. Every point of $f(C) \cap \Theta^{-}_{u}$ can be written

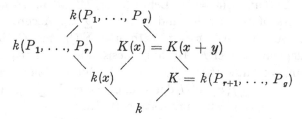

$$f(P) = -\sum_{j=2}^{g} f(P_j) + u$$

with suitable points P, P_j of C. If we put $P = P_1$, we obtain

$$u = \sum_{i=1}^{g} f(P_i).$$

But we know that in such an expression of a generic point of J, the P_i are uniquely determined up to a permutation, and are generic independent. If we put $x = f(P)$ and $y = \sum_{j=2}^{g} f(P_j)$, then we note that $k(x, y)$ is a separable extension of $k(x + y) = k(u)$ because it is contained in the field $k(P_1, \ldots, P_g)$ which is Galois over $k(u)$. Proposition 5 of Chapter I, § 2 now gives us the following result.

PROPOSITION 3. *Let u be a generic point of J. Then*

$$f(C) \cdot \Theta^{-}_{u} = \sum_{i=1}^{g} (f(P_i))$$

where the P_i are independent generic points of C, uniquely determined (up to a permutation) by the relation $u = \sum_{i=1}^{g} f(P_i)$.

We are not going to give here the full theory of the Pontrjagin products of $f(C)$ and of the relations which one can find between these products and their intersections, especially with respect to numerical equivalence. A number of questions concerning this topic are not yet completely cleared up. We shall limit ourselves to proving a result which is sufficient to show that $f(C)$ is biholomorphic to C. Let $P = P_1$ be any point of C, and let $P_2, \ldots, P_g$ be independent generic points of C over $k(P_1)$. Put $v = \sum_{i=1}^{g} f(P_i)$. I contend that the intersection $f(C) \cdot \Theta^{-}_{v}$ is defined. Indeed, let $f(Q)$ be a point of this intersection, and put $Q = Q_1$. We can find $Q_2, \ldots, Q_g$ in C such that

$$\sum_{i=1}^{g} f(Q_i) = \sum_{i=1}^{g} f(P_i).$$

Since the genus of C is > 0, we have $l(P_1) = 1$ (i.e., the space $L(P_1)$ consists only of constants). By an argument entirely analogous to the one of Lemma 5, we conclude, from the fact that $P_2, \ldots, P_g$ are generic, that

$$l(\sum_{i=1}^{g} P_i) = 1.$$

This shows that the Q_i are uniquely determined up to a permutation, and are equal to the P_i. The intersection $f(C) \cdot \Theta^-_v$ is therefore defined. It is obtained by specializing the 0-cycle of Proposition 3, by F—VII$_6$ Th. 13 and Proposition 4 of Chapter I, § 2. Our cycle is of degree g, and all its components have multiplicity 1. A classical criterion (F—VI$_2$ Th. 6) shows that $f(P)$ is simple on $f(C)$, and thus that $f(C)$ is non-singular since P was selected arbitrarily. We therefore obtain the following result.

PROPOSITION 4. *Let $f : C \to J$ be a canonical map of C into its Jacobian. Then all the points of $f(C)$ are simple on $f(C)$ which is biholomorphic to C.*

The above proposition allows us to identify C in its Jacobian. Using Propositions 2, 3, and Proposition 5 of Chapter I, § 2 we have:

COROLLARY. *Suppose that $f : C \to J$ is the inclusion. Then $d(C, \Theta) = g$. If C^{*r} denotes the Pontrjagin product of C with itself r times, then*

$$C^{*r} = r!W_r, \quad C^{*(g-1)} = (g-1)!\,\Theta, \qquad C^{*g} = g!J.$$

Similar arguments can be used to compute $d(W_r, W_{g-r})$. This leads to a finer analysis of the numerical properties of a Jacobian, and eventually to Torelli's theorem, which allows one to recover the curve C once J and the canonical divisor Θ are known. We do not deal with these questions here.

§ 3. *The Albanese variety*

Our purpose in this section is to associate with each variety V an abelian variety A and a rational map of V into A satisfying the universal mapping property for rational maps of V into abelian varieties.

Let $f : V \to A$ be a rational map of V into an abelian variety. We say that (V, f) *generates* A if there exists an integer n such that the map $F : V \times \ldots \times V \to A$ equal to the sum of f with itself n times is generically surjective. After a suitable translation, we may assume that the image of V in A goes through the origin, and in that case one sees that (V, f) generates A if and only if the smallest abelian subvariety of A containing $f(V)$ is equal to A.

An *Albanese variety* (A, f) of V is a couple consisting of an abelian variety A and a rational map $f : V \to A$ such that:

(i) (V, f) generates A;

(ii) for every rational map $g : V \to B$ of V into an abelian variety B, there exists a homomorphism $g_* : A \to B$ and a constant $c \in B$ such that $g = g_* f + c$.

We observe immediately that the homomorphism g_* in (ii) is unique. If we have $g = g'f + c'$, with a homomorphism g', then $g' = g_*$. Indeed, let u be a generic point of A and $P_1, \ldots, P_n$ independent generic points of V such that $u = \sum f(P_i)$. We have $g(P_i) = g_* f(P_i) + c = g' f(P_i) + c'$. Taking the sum, we find

$$g_*(u) = g'(u) + b$$

for some constant $b \in B$. Since g_* and g' are homomorphisms, we have $g_*(0) = g'(0) = 0$, and hence $b = 0$, and $g' = g_*$. We shall say that g_* is the homomorphism *induced* by g.

By an abuse of language, we shall sometimes say that A is an *Albanese variety*, and that f is a *canonical map*.

THEOREM 11. *Let V be a variety. Then there exists an Albanese variety (A, f) of V. The abelian variety A is uniquely determined up to a birational isomorphism, and f is determined up to a translation.*

Proof: The uniqueness of A and f up to a translation is an immediate consequence of the uniqueness of g_* in (ii). To prove the existence, we note that the theorem is birational in V. Hence we may restrict ourselves to an affine variety V.

We are first going to use the generic curve on V to show that the dimension of abelian varieties generated by V is bounded. For the construction of the generic curve, we refer the reader to IAG—VII_6. We shall recall here the properties that we are going to use. If V is defined over k, then the generic curve C_t on V is defined over a regular extension $k(t)$ of k, and t is a generic point of a parameter variety (in fact of a linear variety). If P, Q are two independent generic points of C_t over $k(t)$, then they are independent generic points of V over k. Furthermore, C_t is constructed canonically from the parameters t, so that if t, t' are two generic points of the parameter variety over k, there is an isomorphism $\sigma : k(t) \rightarrow k(t')$ sending t on t', such that $C_t^\sigma = C_{t'}$. We may therefore speak, by abuse of language, of *the* generic curve C_t. Finally, we note that the genus of C_t and $C_{t'}$ (or rather of their normalization) is the same, again because of the isomorphism σ. We may therefore speak of the genus of the generic curve. As usual, we may choose t generic over any given field K containing k. For example, if $f : V \rightarrow B$ is a rational map of V into an abelian variety, defined over K, we shall take C_t generic over K.

The dimension of an abelian variety generated by V is bounded by the genus of the generic curve. This will come from the universal mapping property of the Jacobian, and the following lemma.

LEMMA 6. *If (V, f) generates an abelian variety A, then the restriction of f to the generic curve of V also generates A.*

Proof: Let k be a field of definition for $f : V \rightarrow A$. The generic curve C_t being defined over $k(t)$, let $f_t : C_t \rightarrow A$ be the induced rational map of C_t into A, defined over $k(t)$. We know that if P, Q are two independent generic points of C_t over $k(t)$, then they are independent generic points of V over k. We have a rational map $F : V \times V \rightarrow A$ given by the formula $F(P, Q) = f(P) - f(Q)$, and we denote by $F_t : C_t \times C_t \rightarrow A$ the induced mapping on C_t, given by the same formula. The image of $C_t \times C_t$ in A goes through the origin, and generates an abelian subvariety of A which is defined over k by Chow's theorem (Theorem 5 of § 1). By definition, it contains the point $f(P) - f(Q)$, and hence it contains the image of V in A, which we may suppose goes through the origin. It is therefore equal to A.

If C'_t is the non-singular curve birationally equivalent to C_t, and defined over the algebraic closure of $k(t)$, then the rational map of C_t into A gives rise to a map of C'_t into A which generates

A. Hence there exists a generically surjective homomorphism of the Jacobian of C'_t onto A, and we see that the dimension of A is bounded by the genus of C'_t.

Let now $f : V \to A$ be a rational map of V into an abelian variety. We shall assume, without loss of generality, that the set-theoretic image $f(V)$ goes through the origin. We denote by $f_n : V_n \to A$ the rational map of the product of V with itself n times obtained by taking the sum of f with itself n times. According to what we have just proved, there exists an integer n (at most the genus of the generic curve) such that if (V, f) generates A, then f_n is generically surjective. To prove the existence of the Albanese variety, it will suffice to prove that if k is algebraically closed, then there exists a rational map $f : V \to A$ defined over k satisfying conditions (i) and the universal mapping property (ii) for rational maps of V into abelian varieties *defined over k*, because we may take for k itself a universal domain. In addition, we may always assume that such mappings are normalized: we take a fixed simple point P_0 of V rational over k, and we consider only mappings $f : V \to A$ such that $f(P_0)$ is the origin of A.

If (V, f) generates A, then $k(A) \subset k(V_n)$: the function field of A over k is contained in the function field of the product of V with itself n times. Let $A = A_1$ be an abelian variety of maximal dimension generated by V. Then there cannot exist an infinite tower

$$k(A) = k(A_1) \subset k(A_2) \subset k(A_3) \subset \ldots \subset k(V_n)$$

with proper inclusions, and where each A_i is generated by V. Indeed, the hypothesis that A is of maximal dimension implies that each $k(A_i)$ is algebraic over $k(A_{i-1})$, and since the field $k(V_n)$ is finitely generated over k, and hence over $k(A)$, the algebraic closure of $k(A)$ in $k(V_n)$ is of finite degree over $k(A)$. Hence there exists an abelian variety A and a rational map $f : V \to A$ satisfying the following conditions: (V, f) generates A, and there is no abelian variety generated by V such that $k(A) \subset k(B) \subset k(V_n)$, and such that the inclusion $k(A) \subset k(B)$ is proper. I contend that (A, f) is the desired Albanese variety. To prove this, let $g : V \to B$

be a rational map. We may suppose that (V, g) generates B. Let $h: V \to A \times B$ be the map defined by $h(P) = (f(P), g(P))$. Then $h(P_0) = 0$, and $h(V)$ generates an abelian subvariety A' of $A \times B$. We have obviously

$$k(A) \subset k(A') \subset k(V_n)$$

and hence $k(A) = k(A')$. This means that there exists a birational isomorphism $\lambda: A \to A'$ between A and A' such that, if $P_1, \ldots, P_n$ are independent generic points of V over k, we have

$$\lambda \sum_{i=1}^{n} (f(P_i)) = \sum_{i=1}^{n} h(P_i).$$

If we specialize $P_2, \ldots, P_n$ into P_0, we find $\lambda f(P_1) = h(P_1)$. If α denotes the restriction to A' of the projection of $A \times B$ on B, then we have $\alpha h = g$, and hence $\alpha\lambda f = g$. Thus we have proved the existence of a homomorphism $g_* = \alpha\lambda$ such that $g = g_* f$. This concludes the proof of the existence of the Albanese variety.

We make some remarks concerning the birational and biholomorphic character of the Albanese variety. We work over an algebraically closed field k. Let K be a finitely generated extension of k. Then K is isomorphic to the function field of some variety V, called a model of K over k, if we can write $K = k(P)$ for some generic point P of V. There are of course infinitely many ways of selecting a model for K. Let $f: V \to A$ be a canonical map of V into its Albanese variety, defined over k, and let $f(P) = Q$ be the image of P in A. Let U be the image $f(V)$ of V in A: it is the locus of Q over k. Then no matter what model V we selected for K, the subvariety U of A is uniquely determined up to a biholomorphic correspondence. Thus we could very well speak of a rational map of K into its Albanese variety, and we see that the existence of an Albanese variety in fact allows us to determine in a canonical fashion the following objects associated with the field K:

(a) A subfield E characterized from the birational point of view by the condition that it is the largest subfield of K containing k having a model which is a subvariety of an abelian variety.

(b) A model U of the subfield E which is a subvariety of an abelian variety, and is such that if W is a model of a subfield L of K, and is a subvariety of an abelian variety, then $L \subset K$, and the rational map induced by this inclusion is everywhere defined on U. In other words, U dominates W.

By fundamental principles, it is then clear that every automorphism of K over k induces an automorphism of our subfield E, which may be called the *Albanese subfield* of K. In addition, such an automorphism gives rise to a birational biholomorphic transformation of the model U into itself, and hence it permutes the local rings of U in E.

If we do not work over an algebraically closed constant field, then we must replace the mappings of V into abelian varieties by mappings into principal homogeneous spaces.

We are now going to consider in greater detail the field over which the Albanese variety can be defined. We have seen that the map f is determined only up to a translation. In order to have a completely canonical situation, we proceed as follows. We shall say that a rational map $f : V \times V \to A$ of the product $V \times V$ into an abelian variety is *admissible* if we can find a field k' and a rational map $f' : V \to A$ such that $f(P, Q) = f'(P) - f'(Q)$ for P, Q independent generic points of V over k'. In view of Theorems 2 and 3, this amounts to saying that f vanishes on the diagonal, i.e., $f(P, P) = 0$. Of course, f may be defined over a field k, while f' is not defined over k. We note that f' is uniquely determined by f up to an additive constant.

A couple (A, f) consisting of an abelian variety and an admissible map f of $V \times V$ into A will be called *admissible*. We say that it is *defined over* k if A and f are defined over k.

Let V be a variety defined over k. An admissible couple (A, f) for V will be called a *k-Albanese variety* of V if it is defined over k, and if the following conditions are satisfied:

$(\mathrm{i})_k$ $(V \times V, f)$ generates A;

$(\mathrm{ii})_k$ for every admissible rational map $g : V \times V \to B$ into an abelian variety defined over k, there exists a homomorphism $g_* : A \to B$ defined over k such that $g = g_* f$.

$$\begin{array}{ccc} V \times V & \xrightarrow{f} & A \\ & g \searrow \quad \swarrow g_* & \\ & B & \end{array}$$

As before, we see that the homomorphism g_* is uniquely determined by f and g. In particular, two k-Albanese varieties differ only by a birational isomorphism defined over k. This time, the mapping f is uniquely determined, not only up to an additive constant. We call it the *canonical map*. The set of all k-Albanese varieties (A, f) of V will be denoted by $\mathrm{Alb}_k(V)$. From the biholomorphic point of view, all elements of $\mathrm{Alb}_k(V)$ can be identified.

Let K be a field containing k. If $(A_K, f_K) \in \mathrm{Alb}_K(V)$ and $(A_k, f_k) \in \mathrm{Alb}_k(V)$, then there exists a unique homomorphism

$$I_k^K\colon A_K \to A_k$$

such that $f_k = I_k^K f_K$. We shall see eventually that this homomorphism is always an isomorphism, but that it can be purely inseparable. If it is birational, this means that we can find a K-Albanese variety which is defined over k. By an abuse of notation, we sometimes write $I_k^K : \mathrm{Alb}_K(V) \to \mathrm{Alb}_k(V)$.

The existence of a k-Albanese variety of V is an immediate consequence of the existence of an Albanese variety. If we have two admissible rational maps $f : V \times V \to A$ and $g : V \times V \to B$ defined over k, we can form their supremum, i.e., an admissible map $h : V \times V \to A \times B$ which generates an abelian subvariety C of $A \times B$, and such that there exist two homomorphisms $\alpha : C \to A$ and $\beta : C \to B$ (the projections) such that $f = \alpha h$ and $g = \beta h$. These homomorphisms are *a fortiori* defined over the universal domain, and the existence of the Albanese variety shows that among all admissible maps defined over k, there exists one which is maximal.

The following properties of the k-Albanese variety show its functorial nature.

PROPOSITION 5. *Let $\varphi : V \to W$ be a rational map defined over k, and assume that $\varphi(V)$ is simple on W. Let $(A, f) \in \mathrm{Alb}_k(V)$ and*

$(B, g) \in \mathrm{Alb}_k(W)$. *Then there exists one and only one homomorphism* $\varphi_* : A \to B$ *defined over* k *such that the following diagram is commutative*:

$$\begin{array}{ccc} V \times V & \xrightarrow{\;f\;} & A \\ {\scriptstyle \varphi \times \varphi}\downarrow & & \downarrow{\scriptstyle \varphi_*} \\ W \times W & \xrightarrow[\;g\;]{} & B \end{array}$$

Proof: The mapping $\varphi \times \varphi$ followed by g is defined, according to the hypothesis, and Theorem 2 of § 1. The existence and uniqueness of φ_* are then immediate consequences of the definition of the k-Albanese variety.

In addition to an induced homomorphism over a given field, we also have an induced homomorphism when we change the ground field.

PROPOSITION 6. *Let* V *be a variety defined over a field* K, *and let* W *be a variety defined over* $k \subset K$. *Let* $\varphi : V \to W$ *be a rational map defined over* K *such that* $\varphi(V)$ *is simple on* W. *Let* $(A, f) \in \mathrm{Alb}_K(V)$ *and* $(B, g) \in \mathrm{Alb}_k(W)$. *Then there exists one and only one homomorphism*

$$\varphi_{*k}^{K} : A \to B$$

defined over K *such that the following diagram is commutative*:

$$\begin{array}{ccc} V \times V & \xrightarrow{\;f\;} & A \\ {\scriptstyle \varphi \times \varphi}\downarrow & & \downarrow{\scriptstyle \varphi_{*k}^{K}} \\ W \times W & \xrightarrow[\;g\;]{} & B \end{array}$$

The proof is as obvious as that of the preceding proposition. Our homomorphism φ_* in Proposition 5 is equal to φ_{*k}^{k} in the notation of Proposition 6.

The next formulas are also obvious.

$$\varphi_{*k}^{K} = I_k^K \varphi_{*K}^{K} \tag{1}$$

where I_k^K refers to W. In particular, if $\varphi = \delta$ is the identity on V, then

$$\delta_{*k}^{K} = I_k^K. \tag{2}$$

If U is a variety defined over a field $E \supset K \supset k$, and if $\psi : U \to V$ is a rational map such that $\psi(U)$ is simple on V and such that φ is defined at $\psi(U)$, then we have a composed rational map $\varphi\psi : U \to W$ defined over E, and

$$(\varphi\psi)_{*k}^{E} = \varphi_{*k}^{K} \psi_{*K}^{E}. \tag{3}$$

Finally, we have a particularly important example in characteristic p. The Frobenius automorphism of the universal domain which sends each element onto its p^eth power will be denoted by π_e, or simply π if the integer e is fixed once and for all. It transforms a variety V into a variety $V^{(p^e)}$. This transformation will be denoted by π_V. It is a homomorphism if V is an abelian variety. We have

$$\pi_{*K^{p^e}}^{K} = \pi_A. \tag{4}$$

This remark will be useful to treat the purely inseparable case of the next proposition.

PROPOSITION 7. *Let the notation be the same as in Proposition 5. If φ is purely inseparable (resp. generically surjective) then φ_* is purely inseparable (resp. generically surjective).*

Proof: If φ is purely inseparable, then there exists a rational map $\psi : W \to V^{(p^e)}$ defined over k such that $\psi\varphi = \pi$ for a suitable e. According to (3), we find

$$\pi = (\psi\varphi)_{*k^{p^e}}^{k} = \psi_{*k^{p^e}}^{k} \varphi_{*k}^{k}.$$

Since π is purely inseparable, we see that φ_{*k}^{k} is purely inseparable.

Suppose now that φ is generically surjective, and let (P_i, Q_i) $(i = 1, \ldots, n)$ be independent generic points of $V \times V$ over k such that $u = \sum f(P_i, Q_i)$ is a generic point of A. By hypothesis, $\varphi(P_i, Q_i)$ is a generic point of $W \times W$ by hypothesis, and $\sum g\varphi(P_i, Q_i) = v$ is a generic point of B, provided we have taken n

sufficiently large. We have

$$v = \sum g\varphi(P_i, Q_i) = \sum \varphi_* f(P_i, Q_i) = \varphi_* u.$$

Hence φ_* is generically surjective.

PROPOSITION 8. *Let $V = \prod V_i$ be a product of varieties defined over k. If $(A_i, f_i) \in \mathrm{Alb}_k(V_i)$ then $(\prod A_i, \prod f_i)$ is a k-Albanese variety of $\prod V_i$.*

Proof: Obvious, in view of Theorem 3, § 1.

PROPOSITION 9. *Let V be a variety defined over k, and let $K \supset E \supset k$ be two extensions of k. Let (A, f) be a K-Albanese variety of V, and assume that (A, f) is defined over E. Then $(A, f) \in \mathrm{Alb}_E(V)$.*

Proof: Let $(B, g) \in \mathrm{Alb}_E(V)$. There exists a homomorphism $\varphi_* : B \to A$ defined over E, arising from the E-universality of (B, g), such that $f = \varphi_* g$. It is the homomorphism induced by the identity, relative to the field E. There exists a homomorphism $g_* : A \to B$ defined over K such that $g = g_* f$. It is the homomorphism induced by g relative to the K-universality of (A, f). We get $f = \varphi_* g_* f$. Over the field K, we must have $f_* = \delta_A = \varphi_* g_*$, and hence φ_* is birational. This proves that $(A, f) \in \mathrm{Alb}_E(V)$.

Proposition 9 shows in particular that if A is an Albanese variety of V relative to the universal domain, if $f' : V \to A$ is one of the canonical mappings, and if f' is defined over a field E, then (A, f) is an E-Albanese variety of V, if we put $f(P, Q) = f'(P) - f'(Q)$.

We are led to study the behaviour of $\mathrm{Alb}_K(V)$ for the three cases where K is a regular, separable algebraic, and purely inseparable extension of k. The first two cases constitute the separable case.

THEOREM 12. *Let V be a variety defined over a field k. Let K be an extension of k, and $I_k^K : \mathrm{Alb}_K(V) \to \mathrm{Alb}_k(V)$ the canonical homomorphism. If K is purely inseparable over k, then I_k^K is a purely inseparable isomorphism. If K is separable over k, then I_k^K is a birational isomorphism.*

Proof: Using Proposition 9 we are immediately brought back to the consideration of a finitely generated extension K of k.

Suppose first that K is purely inseparable and finite over k. There exists a power p^e of the characteristic such that K^{p^e} is contained in k. In view of (4), and of (1), (2), (3) we find

$$\pi = (\pi\delta)_{*K^{p^e}}^{K} = \pi_{*K^{p^e}}^{k} \delta_{*k}^{K}$$

and hence $\delta_{*k}^{K} = I_k^K$ is purely inseparable.

In order to treat the separable case, we are going to use the general theorems concerning the field of definition of a group variety recalled in Chapter 1, § 3.

Let K be a separable extension of k, of infinite transcendence degree, and separably algebraically closed, i.e., equal to its separable algebraic closure. We can say that K is a universal domain for separable extensions of k. In order to prove Theorem 12, it will suffice to prove that we can find a K-Albanese variety of V which is defined over k. Indeed, Proposition 9 then garantees that this Albanese variety is also an E-Albanese variety for every separable extension E of k between k and K.

We are going to deal separately with regular extensions and separable algebraic extensions. We suppose at first that k is equal to its separable algebraic closure k_s. Our field K is then a regular extension of k, and a K-Albanese variety (A_K, f_K) in $\mathrm{Alb}_K(V)$ is defined over a finitely generated extension $k(t)$ of k. We can thus write $A_K = A_t$, $f_K = f_t$, the index t indicating the field of definition.

Let t' be a generic specialization of t, such that $k(t)$ and $k(t')$ are independent over k, and $k(t, t')$ is contained in K. The isomorphism of $k(t)$ onto $k(t')$ sending t on t' transforms (A_t, f_t) into $(A_{t'}, f_{t'})$. This isomorphism can be extended to an automorphism of K over k. By transport of structure, it follows that $(A_{t'}, f_{t'}) \in \mathrm{Alb}_K(V)$, and we note that $(A_{t'}, f_{t'})$ is defined over $k(t')$.

By Proposition 9, we know that (A_t, f_t) is an E-Albanese variety of V for every field E between $k(t)$ and K. A similar remark applies to $(A_{t'}, f_{t'})$. In particular, take $E = k(t, t')$. By definition, there exist homomorphisms

$$g_{t,t'} : A_{t'} \to A_t, \qquad g_{t',t} : A_t \to A_{t'}$$

defined over $k(t, t')$ such that the following diagrams are commutative:

$$\begin{array}{ccc} & V\times V & \\ {}^{f_t}\swarrow & & \searrow^{f_{t'}} \\ A_t & \xrightarrow[g_{t',t}]{} & A_{t'} \end{array} \qquad \begin{array}{ccc} & V\times V & \\ {}^{f_t}\swarrow & & \searrow^{f_{t'}} \\ A_t & \xleftarrow[g_{t,t'}]{} & A_{t'} \end{array}$$

These homomorphisms are uniquely determined, and are inverse to each other. They are therefore birational isomorphisms. We see immediately that the coherence conditions of Chapter I, § 3 Theorems 1, 1G are satisfied. Hence there exists an abelian variety A defined over k, a birational isomorphism $g_t : A \to A_t$ defined over $k(t)$, and a rational map $f : V \times V \to A$ also defined over k such that the following diagram is commutative:

$$\begin{array}{ccc} V\times V & \xrightarrow{f_t} & A_t \\ {}^{f}\searrow & & \nearrow_{g_t} \\ & A & \end{array}$$

The map f is equal to $g_t^{-1}f_t$. This shows that f is admissible, i.e., vanishes on the diagonal, and concludes the proof in case K is a regular extension of k.

We return to an arbitrary field k, and still use the universal separable domain K of k. We have just shown that we can find a K-Albanese variety of V which is defined over the separable algebraic closure k_s of k. It is in fact defined over a finite separable extension k_1 of k. The automorphisms of k_s which leave k fixed will be denoted by σ, τ. The K-Albanese variety (A_K, f_K) can be written (A_1, f_1) since it is defined over k_1. Let $(A_1{}^\sigma, f_1{}^\sigma)$ be the transform of (A_1, f_1) by σ. It is also a k_s-Albanese variety of V, by transport of structure. Hence there exist birational isomorphisms

$$g_{\sigma,\tau} : A_1{}^\tau \to A_1{}^\sigma$$

satisfying the commutativity relation $f_1{}^\sigma = g_{\sigma,\tau} f_1{}^\tau$ and the coherence conditions of Chapter I, § 3 Theorems 2, 2G, because of the uniqueness of such isomorphisms. Hence we can find an abelian variety A defined over k, and a birational isomorphism $g_\sigma : A \to A_1{}^\sigma$

such that $g_{\sigma,\tau} g_\tau = g_\sigma$. If we put $f = g_\sigma^{-1} f_1{}^\sigma$ then we see that f is defined over k, and is admissible. We have thus shown that (A, f) is a k-Albanese variety birationally isomorphic to (A_K, f_K) over K. This concludes the proof of Theorem 12.

REMARK. We shall see in Chapter VII that if a function field over k has a projective model which is non-singular in codimension 1, then the isomorphism I_k^K is birational in all cases. This will be shown by induction on the dimension of V, the theorem being true for non-singular curves, in view of the fact that such a curve always has a simple point which is rational over a separable extension of a given field of definition.

Historical Note:

The fundamental theorems concerning rational maps of varieties into abelian varieties come from Weil [85], and we have reproduced his theory almost without change. We have also followed his construction for the Jacobian variety.

Matsusaka [53] was the first one to give a construction for the Albanese variety, using the generic curve. Serre [76] has shown how one can avoid using the Jacobian in order to bound the dimension of abelian varieties generated by a given variety, by using differential forms.

As we have already mentioned, Chow [17] had constructed the Jacobian of a curve over the given field of definition of the curve, and while Matsusaka obtained the Albanese variety only over the universal domain, Chow [15] gives a construction based on his theory of algebraic systems of abelian varieties [14] which gives the Albanese variety over the given ground field. In substance, it is his proof which we have given here, except that we have used the criteria concerning the lowering of a field of definition in the form given to them by Weil [92]. As Weil himself noticed, this short-circuits certain portions of Chow's theory.

Chow's projective construction of the Jacobian is of considerable importance because, roughly speaking, it does not introduce denominators, i.e., it belongs to absolute geometry. Here of course, we have worked birationally and biholomorphically, and we have completely lost track of the integral nature of parameters on which the curve may depend. On the other hand, Chow's method gives rise to Igusa's compatibility principle which allows us to study algebraic systems of Jacobian varieties [39], [40]. For instance, if we are given an algebraic system of curves, together with their canonical mappings into their Jacobians, if (C, J, f) is a generic member, and if C' is a specialization of C which is non-singular, then (J, f) has a uniquely determined specialization (J', f') which is the Jacobian variety of C'. Here of course specialization may include reduction mod p. In addition, Igusa has also worked out what happens when C' acquires certain singularities, in which case (J', f') becomes a generalized Jacobian of Rosenlicht [70], [71].

These generalized Jacobians provide an analysis of a curve with singularities both from the point of view of the universal mapping property of the curve into commutative algebraic groups [74] and from the point of view of divisor classes. They can be used to classify the abelian coverings of the curve [43], [47]. Their analogues in higher dimension are still lacking.

CHAPTER III

The Theorem of the Square

In the first section, we recall the notion of algebraic equivalence, analogous to the notion of homotopy in topology, except that we require all objects entering into its definition to be algebraic, so that we really end up with an equivalence which is more like that of homology.

The theorem of the square which follows furnishes the key to the entire theory of divisor classes on an abelian variety, modulo various equivalences (numerical, algebraic, torsion, linear, etc.) to be studied in the next chapter. The fact that Weil was able to give a direct proof, using only intersection theory and the existence of the Jacobian, will allow us to develop the rest of the theory without pains.

After stating and proving the theorem of the square, we derive some useful corollaries concerning its application to algebraic groups. In particular, we shall see that if X is a positive divisor on an abelian variety A, then the set of points $a \in A$ such that X_a is linearly equivalent to X is an algebraic subgroup of A. This is the first step in the construction of the Picard variety, given in the next chapter.

§ 1. *Algebraic equivalence*

Let V, W be varieties (abstract, as usual), and let Z be a cycle on the product $V \times W$. We shall always denote by $Z(M)$ the cycle on W defined by the formula

$$M \times Z(M) = Z \cdot (M \times W)$$

whenever M is a simple point of V such that $Z \cdot (M \times W)$ is defined. If N is a simple point of W such that $Z \cdot (V \times N)$ is defined, we denote by ${}^tZ(N)$ the cycle on V such that

$$ {}^tZ(N) \times N = Z \cdot (V \times N). $$

Thus ${}^tZ$ is the transpose of Z on $W \times V$. It will play an important role in the duality theory of abelian varieties, but we shall not make use of it for the rest of this chapter.

Let X be a component of Z, and X' its projection on V. If $Z \cdot (M \times V)$ is defined, then $X \cdot (M \times W)$ is also defined. But, according to the dimension theorem for correspondences (F-VII$_4$ Prop. 8) every component of $X \cap (M \times W)$ has at least the dimension dim (X) — dim (X'). If it is simple on $V \times W$, it can be proper only if it is of dimension dim (X) — dim (W). Thus we must have $X' = V$. Let Z_0 be the cycle whose components are those components of Z having projection V on V, taken with coefficients equal to those which they have in Z. Then $Z_0(M)$ is defined whenever $Z(M)$ is defined, and we have $Z_0(M) = Z(M)$. Hence whenever we deal with cycles of type $Z(M)$, we may assume, without loss of generality, that all the components of Z have projection V on V. This implies of course that the dimension of Z is at least equal to that of V.

The cycles $Z(M)$ are interpreted geometrically as members of an algebraic family, parametrized by V. If M is a generic point of V (over some field of definition of V, W over which Z is rational) then $Z(M)$ is a generic member of this family.

We shall say that a cycle X on W is *algebraically equivalent to zero*, and write $X \approx 0$, if there exists a variety V (called a *parameter variety*), a cycle Z on $V \times W$, and two simple points M, N on V such that $Z(M)$ and $Z(N)$ are defined and such that

$$ X = Z(M) - Z(N). \tag{1} $$

We are going to give some results which will make it easier for us to handle algebraic equivalence, and in Proposition 1, we shall give properties of algebraic equivalence which could equally well have been taken as definitions. In particular, we shall prove that for algebraic equivalence, we might have assumed that the parameter variety is a curve, or an abelian variety (in fact, the Jacobian of a curve). Intuitively, we reason as follows. We take a curve through the points M, N on V, and we induce the cycle Z on the product of this curve with W. This shows that we may assume V to be a curve. We can even suppose that it is complete, non-singular, because the parametrization is local in M, N. Finally, we use the existence of the Jacobian and the fact that a complete non-singular curve can be embedded biholomorphically in its Jacobian to lift the parametrization to the Jacobian.

The purpose of Lemmas 1, 2 and Proposition 1 is to formalize the preceding heuristic arguments. The reader who wishes to get the main ideas of this chapter should of course skip them.

LEMMA 1. *Let V, W be varieties, V' a simple subvariety of V, and Z a cycle on $V \times W$. Let Z_h be the components of Z, and let a_h be their coefficients in Z. Let Z'_{hj} be all the distinct proper components of $Z_h \cap (V' \times W)$ having projection V' on V. Put*

$$Z' = \sum_{h,j} i(Z_h \cdot (V' \times W), Z'_{hj}; V \times W) a_h Z'_{hj}.$$

Then Z' is a cycle on $V' \times W$. If k is a field of definition for V, W, V' over which Z is rational, then Z' is rational over k. If M is a point of V', which is simple on both V and V' such that $Z(M)$ is defined, then $Z'(M)$ is also defined, and is equal to $Z(M)$.

Proof: We note that in the statement of the theorem, $Z(M)$ and $Z'(M)$ are defined, respectively, by the relations

$$M \times Z(M) = \{Z \cdot (M \times W)\}_{V \times W},$$
$$M \times Z'(M) = \{Z' \cdot (M \times W)\}_{V' \times W}.$$

One sees immediately from F-IV$_2$ Th. 4 that Z' is rational over k. For the remaining parts of the theorem, it will suffice, by linearity, to give the proof when Z is a variety. If the projection of Z on V is not equal to V, then $Z \cdot (M \times W) = 0$ when it is defined. If some component Z'_j of $Z \cap (V' \times W)$ has projection V' on V, then V' must be contained in the projection of Z on V. But then the dimension theorem for correspondences (F-VII$_4$ Prop. 8) shows that Z'_j cannot be proper. Hence in this case, we have $Z' = 0$ and our lemma is proved. We may therefore assume that Z has projection V on V.

We are going to show that whenever M is a point of V', simple on both V and V', such that $Z \cdot (M \times W)$ is defined on $V \times W$, then $Z' \cdot (M \times W)$ is defined on $V' \times W$, and these two cycles have the same components. Once this has been done, our lemma will follow by applying F-VI$_3$ Th. 9 to $V \times W$, $V' \times W$, $M \times W$, and Z.

Put $m = \dim V$, $m' = \dim V'$, and $m + r = \dim Z$. Assume that M is a simple point of V and V' such that $Z \cdot (M \times W)$ is defined.

Let X be a component of $Z \cdot (M \times W)$, i.e., a component of $Z \cap (M \times W)$ which is simple on $V \times W$. Then $\dim X = r$. Let T be a component of $Z \cap (V' \times W)$ containing X. Then T is simple on $V \times W$, and according to the dimension theorem for correspondences, its dimension s is $\geqq m + r$. If Y is a component of $T \cap (M \times W)$ containing X, then Y is contained in Z and in $M \times W$, so that $X = Y$. Consequently, X is a component of $T \cap (M \times W)$. We are going to prove that T is one of the Z'_j. Indeed, let T' be the projection of T on V and t' its dimension. Still by the dimension theorem, we have $\dim X \geqq s - t'$, and thus $r \geqq s - t'$, whence $t' \geqq s - r \geqq m'$. But $T' \subset V'$, and so $t' \leqq m'$, whence $t' = s - r = m'$ and $T' = V'$. Consequently T is one of the Z'_j. Since M has been assumed simple on V', we know that X is simple on $V' \times W$ (F-IV$_6$ Th. 13, Cor. 1) and is a proper component of $T \cap (M \times W)$ on $V' \times W$. Every component of $Z \cdot (M \times W)$ is, under the above conditions, a proper component of one of the intersections $Z'_j \cap (M \times W)$ on $V' \times W$.

Conversely, let T be one of the Z'_j. Then T is simple on $V \times W$ and hence has a projection on W which is simple on W. Hence T is simple on $V' \times W$ (F-IV$_4$ Th. 13, Cor. 1), and we have already shown that Z' is a cycle on $V' \times W$. Let Y be a component of $T \cap (M \times W)$ simple on $V' \times W$. Since T is a proper component of $Z \cap (V' \times W)$ on $V \times W$ and $\dim T = m' + r$, we see that $\dim Y \geqq r$. Furthermore, Y is simple on $V \times W$ (always according to F-IV$_4$ Th. 13, Cor. 1). Hence every component of $Z \cap (M \times W)$

containing Y is simple on $V \times W$. Since $Z \cdot (M \times W)$ is defined, it is proper, and hence is of dimension r. Again from the hypothesis that $Z \cdot (M \times W)$ is defined, we conclude that Y is a component of this intersection; we have also shown that in this case, $\{T \cdot (M \times W)\}_{V' \times W}$ is defined. This concludes the proof of the assertions made above, which allowed us to apply F-VI_3 Th. 9, and therefore the proof of the lemma.

LEMMA 2. *Let W be a variety defined over a field k, P a point of W, and M a generic point of W over $k(P)$. Then there exists a curve C on W passing through M and P, such that M is simple on C. If in addition P is simple on W, then there exists such a curve C on which P is also simple.*

Proof: Since M is a generic point, there exists an affine representative of W on which both P and M have representatives. We may thus assume without loss of generality that W is affine. Put $K = k(P)$. After an affine translation, defined over K, we need only consider the case where P is the origin on the ambient affine space S^n of W. Let $u_1, \ldots, u_n$ be algebraically independent quantities over K, and let L_u be the hyperplane $u_1X_1 + \ldots + u_nX_n = 0$. If P is simple on W, then L_u is transversal to W at P, and hence the intersection $W \cdot L_u$ contains one and only one component going through 0, of dimension $\dim W - 1$, and P is simple on this component. (We have of course assumed that $\dim W \geqq 2$.) According to the dimension theorem, if T is a component of $W \cdot L_u$ passing through 0, then $\dim T = \dim W - 1$. As for the generic hyperplane section, we are going to show that every generic point (x) of T over the algebraic closure $\overline{K(u)}$ of $K(u)$ is a generic point of W over K. We observe that the dimension of (x) over $K(u)$ is equal to $r - 1$, if we put $r = \dim W$. On the other hand, if we assume for instance that $x_1 \neq 0$, we have

$$u_1 = - \sum_{i=2}^{n} u_i x_i / x_1.$$

This shows that the dimension of $K(u, x)$ over $K(x)$ is $\leqq n - 1$. Since we have $\dim_K (x, u) = n + r - 1$ we conclude that $\dim_K(x) = r$, and hence that (x) is a generic point of W over K. Since M is also generic over K, there exists an isomorphism σ of $K(x)$ onto $K(M)$ mapping (x) on M, and leaving $\bar{K}$ fixed. Replacing T by T^σ we are thus reduced to proving our lemma for a variety of dimension smaller than that of W. The proof can be finished by induction.

We are now in a position to prove the results already announced above concerning the parametrization of cycles algebraically equivalent to 0. Let W be a variety of dimension r, and let s be an integer $\leqq r$. In all the applications to be made later, we shall have $s = r - 1$. Denote by $Z_a(W)$, $Z'_a(W)$, $Z''_a(W)$ and $Z^*_a(W)$ the four sets of cycles of dimension s on W, defined as follows:

Z_a: The set of cycles of dimension s algebraically equivalent to 0.

Z'_a: The set of cycles of dimension s obtained by formula (1) when we require that the parameter variety is a curve.

Z''_a: The set of cycles of dimension s obtained by formula (1) when we require the parameter variety to be the Jacobian of a curve.

Z^*_a: The subgroup of the additive group of cycles of dimension s on W which is generated by the elements of Z_a.

PROPOSITION 1. *With the above notations, the four sets of cycles* Z_a, Z'_a, Z''_a, Z^*_a *are equal.*

Proof: Let us first show that Z^*_a is generated by the elements of Z'_a, or in other words that Z_a is contained in the group generated by Z'_a. Indeed, let $Z(N) - Z(M)$ be an element of Z_a, the cycle Z being on a product $V \times W$, and M, N being two simple points of V. Let k be a field of definition for V, W over which Z is rational. Let P be a generic point of V over $k(M, N)$. According to F-VII$_6$ Th. 12, (i), $Z(P)$ is defined. Since we have

$$Z(N) - Z(M) = Z(N) - Z(P) + Z(P) - Z(M)$$

it will suffice, to prove our assertion, to show that $Z(M) - Z(P)$ is in Z'_a, because then we shall also see that $Z(N) - Z(P)$ is in Z'_a. According to Lemma 2, we can find on V a curve C passing through M and P such that M, P are simple on C. According to Lemma 1 there exists on $C \times V$ a cycle Z' such that $Z'(M) = Z(M)$ and $Z'(P) = Z(P)$. This shows that $Z(M) - Z(P) \in Z'_a$.

Our next step is to show that Z'_a is a group. According to the preceding discussion, it will follow that $Z_a^* = Z'_a = Z_a$. Let C_1, C_2 be two curves, Z_1^{s+1} a cycle on $C_1 \times W$, and Z_2^{s+1} a cycle on $C_2 \times W$, M_1 and N_1 two simple points of C_1, and M_2, N_2 two simple points of C_2 such that $Z_1(M_1)$, $Z_1(N_1)$, $Z_2(M_2)$, $Z_2(N_2)$ are defined. Put

$$X_1 = Z_1(N_1) - Z_1(M_1), \text{ and } X_2 = Z_2(N_2) - Z_2(M_2).$$

There exists on $C_1 \times C_2$ a curve D passing through $M_1 \times M_2$ and $N_1 \times N_2$. (The curves C_i can be embedded in projective space, the surface $C_1 \times C_2$ also. The existence of D is then trivial.) There exists a non-singular curve E and a birational map $f : E \to D$ of E onto D, so that we can write $D = f(E)$. We can then find two points P, Q on E such that $f(P) = M_1 \times M_2$ and $f(Q) = N_1 \times N_2$. Let Y_2 be the cycle $E \times C_1 \times Z_2$ on $E \times C_1 \times C_2 \times W$ and let Y_1 be the analogous cycle with respect to C_1, i.e., the cycle on $E \times C_1 \times C_2 \times W$ obtained from $E \times C_2 \times Z_1$ by permuting the factors C_1 and C_2. We obviously have

$$Y_1 \cdot (P \times M_1 \times M_2 \times W) = P \times M_1 \times M_2 \times Z_1(M_1),$$
$$Y_2 \cdot (P \times M_1 \times M_2 \times W) = P \times M_1 \times M_2 \times Z_2(M_2),$$

and we have analogous formulas replacing M_1, M_2 by N_1, N_2. If we put $T = Y_1 - Y_2$, then we find

$$X_1 - X_2 = T(Q \times N_1 \times N_2) - T(P \times M_1 \times M_2).$$

Let Γ be the graph of f. It is a curve on the variety $E \times C_1 \times C_2$ which has become a new parameter variety. According to Lemma 1 there exists a variety T' on $\Gamma \times W$ such that the two terms on the right-hand side of the preceding expression are equal, respectively, to $T'(Q \times N_1 \times N_2)$ and $T'(P \times M_1 \times M_2)$. This shows that $X_1 - X_2 \in Z'_a$ and proves that Z'_a is a group.

Since Z'_a is contained in Z_a, it follows that $Z'_a = Z_a = Z_a^*$. There remains to be shown that $Z'_a \subset Z''_a$, or in other words that we can always take an abelian variety for the parameter variety.

In order to do this, let C be a curve, Z^{s+1} a cycle on $C \times W$, and M, N two simple points of C such that $Z(M)$ and $Z(N)$ are defined. Put $X = Z(N) - Z(M)$. After having replaced C by a curve which is birationally equivalent to it, we may assume without loss of generality that C is complete, non-singular. Let g be its genus, and assume first that $g \neq 0$. Let J be the Jacobian of C, and $f : C \to J$ a canonical map of C into J. Let k be a field of definition for W and f (hence for C, J according to our conventions) over which Z is rational. Let $M_1, \ldots, M_g$ be g independent generic points of C over $k(M, N)$. and put

$$u = \sum_{i=1}^{g} f(M_i).$$

We have $k(u) = k(M_1, \ldots, M_g)_s$. Let

$$\mathfrak{m} = \sum_{i=1}^{g} M_i.$$

It is a divisor on C, rational over $k(u)$. Since

$$Z(M_i) = \mathrm{pr}_W [Z \cdot (M_i \times W)],$$

we get

$$\sum_{i=1}^{g} Z(M_i) = \mathrm{pr}_W[Z \cdot (\mathfrak{m} \times W)].$$

The first term is therefore a rational cycle over $k(u)$, and according to F-VII$_6$ Th. 12 (iii), there exists a cycle Y^{s+g} on $J \times W$ rational over k all of whose components have projection J on J, and such that $Y(u) = \sum Z(M_i)$.

Put $u_0 = \sum_{j=2}^{g} f(M_j)$, $f' = f + u_0$, and $C' = f'(C)$. It is a translation of the image of C in its Jacobian, and hence f' gives a birational biholomorphic correspondence between C and C' defined over the field $k(M_2, \ldots, M_g)$, according to Proposition 4 of Chapter II, § 2. According to Lemma 1, there exists a cycle Y' on $C' \times W$ such that

$$Y'[f'(P)] = Y[f'(P)]$$

for any point P of C such that $Y[f'(P)]$ is defined, and we may assume that all the components of Y' have projection C' on C' (cf. the remarks preceding Lemma 1). In particular, we have $f'(M_1) = u$, and hence

$$Y'[f'(M_1)] = Z(M_1) + \sum_{j=2}^{g} Z(M_j).$$

Since M_1 is generic on C over $k(M_2, \ldots, M_g)$, this implies that after identifying C and C' by means of f', the two cycles Y' and $Z + \{C \times \sum_{j=2}^{g} Z(M_j)\}$ differ only by components whose projection on C is not equal to C, i.e., by degenerate components (cf. F-VII$_6$ Th. 12, (ii)). Consequently, since $Z(M)$ is defined by hypothesis, we get

$$Y'[f'(M)] = Z(M) + \sum_{j=2}^{g} Z(M_j),$$

and similarly for $Y'[f'(N)]$. From this it follows that

$$X = Z(N) - Z(M) = Y'[f'(N)] - Y'[f'(M)]$$

and thus finally that $X = Y[f'(N)] - Y[f'(M)]$ provided that the two terms in the right-hand side are defined.

In order to verify this last point, observe that the locus V of the point

$$v = f'(M) = f(M) + \sum_{j=2}^{g} f(M_j)$$

with respect to the field $K = k(M)$ has dimension $g - 1$. Suppose that Y has a component Y_0 such that $Y_1 \cdot (v \times W)$ is not defined. Let $v \times Q$ be a generic point of a component of $Y_0 \cap (v \times W)$ of dimension $> s$ over the algebraic closure of $K(v)$. Then the dimension of Q over $K(v)$ is $\geqq s + 1$, and the locus of $v \times Q$ over $\bar{K}$ will be a subvariety of Y_0 of dimension $\geqq s + g$. This locus is therefore equal to Y_0, and hence Y_0 has projection V on J, contradicting the definition of Y. Similarly, $Y[f'(N)]$ is defined, and hence we have concluded the proof that $X \in Z''_a$ in the case $g \neq 0$.

Suppose lastly that $g = 0$. Then C is the projective line. Let A be a curve of genus 1, which we identify with its Jacobian. Let $f : A \rightarrow C$ be a generically surjective rational map, and let k be a field of definition for f, W over which Z is rational. Let Γ be the graph of f in $A \times C$. Let P, Q be points of A such that $f(P) = M$ and $f(Q) = N$. Let T be the cycle $A \times Z$ on $A \times C \times W$. It is clear that

$$T(P \times M) = Z(M), \quad \text{and} \quad T(Q \times N) = Z(N).$$

According to Lemma 1, there exists a cycle T' on $\Gamma \times W$ such that

$$T'(P \times M) = T(P \times M), \quad \text{and} \quad T(Q \times N) = T'(Q \times N),$$

whence

$$X = T'(Q \times N) - T'(P \times M).$$

Since we may identify Γ with A by means of the projection of Γ on A, which is a birational biholomorphic correspondence between Γ and A, we see that Proposition 1 is completely proved.

We conclude this section with a description of the other equivalence relations which we may impose on cycles, and restrict ourselves to cycles of codimension 1, i.e., divisors.

As before, let W be a variety. A divisor X on W is said to be *linearly equivalent to* 0 and we write $X \sim 0$ if X is the divisor of a function (i.e., of a rational map of W into the projective line) other than the constant 0 or ∞. If φ is such a function, its divisor (φ) can be written

$$(\varphi) = \varphi^{-1}[(0) - (\infty)]$$

and this divisor is algebraically equivalent to 0. More generally, it follows from F-VIII$_2$ Th. 5 that if D is the projective line and Z is a divisor on $D \times W$, then $Z(M) - Z(N) \sim 0$ whenever M, N are points of D such that $Z(M)$ and $Z(N)$ are defined.

The groups of divisors algebraically equivalent to 0 and linearly equivalent to 0 on a variety W will be denoted by $D_a(W)$ and $D_l(W)$ respectively. The factor group $D_a(W)/D_l(W)$ will be called the *Picard group* of W, and will be denoted by $\mathrm{Pic}(W)$. We note that we have made no assumptions concerning the absence of singularities on W, nor have we assumed that W is complete. If W is complete and non-singular in codimension 1, we shall prove in the next chapter that Pic (W) can be given the structure of an abelian variety.

If X is a divisor on a variety W, we denote by $\mathrm{Cl}(X)$ its linear equivalence class. In the case where W is complete, and non-singular in codimension 1, we shall eventually define $\mathrm{Cl}(X)$ to be a point on the Picard variety, when X is algebraically equivalent to 0.

Denoting by $D(W)$ the group of all divisors on W, the factor group $D(W)/D_a(W)$ is called the *group of Néron-Severi*. If W is complete and non-singular in codimension 1, then it can be shown that it has only a finite number of generators. We shall prove this fact in this book only for abelian varieties.

Let $f : U \to V$ be a rational map. Assume that V is complete and non-singular. Using Corollary 3 of Theorem 3, in the Appendix, § 1, and Proposition 4 of the Appendix, § 2, we can define a homomorphism

$$f^{-1} : D(V)/D_l(V) \to D(U)/D_l(U)$$

by taking the inverse image of a divisor in a linear equivalence class on V. As an immediate consequence of Theorem 2 of the Appendix, § 1, we get:

PROPOSITION 2. *Let U, V, W be three varieties and assume that V, W are complete, non-singular. Let $f : U \to V$ and $g : V \to W$ be two rational maps, everywhere defined. Let $h : U \to W$ be the composed rational map gf. Then we have $(gf)^{-1} = f^{-1}g^{-1}$ for the induced homomorphisms on the linear equivalence classes.*

The analogous proposition for algebraic equivalence is best considered when we come to the transpose of a homomorphism on the Picard variety, in Chapter V.

Let V be a variety, complete and non-singular in codimension 1. A divisor X on V is said to be a *torsion divisor* if there exists an integer $m \neq 0$ such that $m \cdot X$ is algebraically equivalent to 0. The set of all such divisors is a group, denoted by $D_\tau(V)$. The factor group $D_\tau(V)/D_a(V)$ is called the *torsion group* (of divisor classes). We shall study the torsion group of an abelian variety later, and prove that its order is a power of the characteristic. Barsotti has shown that actually the torsion group is trivial: If X is a divisor such that $m \cdot X \approx 0$ with $m \neq 0$, then $X \approx 0$. This fact will not be used in this book.

It can be shown [24] that for the complete non-singular models of a function field, the torsion group is a birational invariant, and that its order is finite (it is a subgroup of the Néron-Severi group). This order has been called classically the number σ of Severi. It is not a birational invariant if we allow complete models, non-singular in codimension 1, but having other singularities. For instance, on an abelian variety A, take the group of automorphisms of order 2 generated by the map $u \to -u$. Its quotient variety W has only a finite number of singularities, coming from the fixed points under the map $u \to -u$: they are the points of order 2 on A. Considering only the classical case where A is a torus (topologically) we see that there are exactly 2^{2r} such points. The covering of W by A is ramified only in these points, and if $\dim A \geqq 2$, the covering is unramified over the divisors of W. This implies that there must be torsion of order 2 in the torsion group of W. However, if we desingularize W, we get a variety whose torsion group is equal to 0 (Spanier [79]).

The above example also shows that if $f: U \to V$ is a rational map, U and V being assumed to be complete and non-singular in codimension 1, and if Y is a divisor on V which is algebraically equivalent to 0 on V, then it may very well happen that $f^{-1}(Y)$ is defined, but not algebraically equivalent to 0 on U; for, otherwise one could show the birational invariance of the torsion group for such varieties, which we have just seen is false.

A similar remark holds for linear equivalence: The absence of certain singularities is essential for the validity of the statement that $f^{-1}(Y) \sim 0$ when $Y \sim 0$, even when f is an inclusion mapping. For instance, we recall the following important result of *Foundations* (F-VIII$_2$ Th. 4, Cor. 1), which we state here as a separate proposition, to be used frequently in the sequel.

PROPOSITION 3. *Let U be a variety, U' a subvariety of U such that every subvariety of U' of codimension 1 on U' is simple both on U and U'. Let φ be a function on U, and assume that φ induces a function φ' on U' which is neither the constant 0 nor ∞. Then*

$$(\varphi') = (\varphi) \cdot U'.$$

Here of course, (φ') denotes the divisor of φ' on U', while (φ) denotes the divisor of φ on U. As an important special case, we get:

COROLLARY. *Let X be a divisor on a product $U \times V$ of two arbitrary varieties and assume $X \sim 0$. If P is a simple point of U such that $X(P)$ is defined, then $X(P) \sim 0$ on V.*

Proof: The subvariety $P \times V$ of $U \times V$ satisfies the hypothesis of Proposition 3, and the projection on V establishes a biholomorphic correspondence between $P \times V$ and V. Our corollary is therefore obvious from the proposition.

§ 2. *The theorem of the cube and the theorem of the square*

Let U, V, W be three varieties defined over a field k, and let X be a divisor on the product $U \times V \times W$, rational over k. Take two copies of U, V, W which we denote by U_i, V_i, W_i $(i = 0, 1)$ and form the product

$$U_0 \times U_1 \times V_0 \times V_1 \times W_0 \times W_1 = U^{(2)} \times V^{(2)} \times W^{(2)}.$$

On this product, we shall denote by X_{ijk} the divisor which is the image of $X \times U \times V \times W$ under the biholomorphic transformation sending the first three factors of $U \times V \times W \times U \times V \times W$ on $U_i \times V_j \times W_k$ and sending the other three on the other three copies of U, V, W. We may say that X_{ijk} is simply X on the product $U_i \times V_j \times W_k$ and the full variety on the other three factors.

Let (u_0, u_1, v_0, v_1) be a generic point of $U^{(2)} \times V^{(2)}$ over k. Consider the cycle

$$Z = \sum (-1)^{i+j+k} X_{ijk}$$

on $U^{(2)} \times V^{(2)} \times W^{(2)}$, and the cycle

$$\begin{aligned} Y &= X(u_0, v_0) - X(u_0, v_1) - X(u_1, v_0) + X(u_1, v_1) \\ &= \sum (-1)^{i+j} X(u_i, v_j) \end{aligned}$$

on W. Then Z is rational over k, and Y is rational over $k(u_0, u_1, v_0, v_1)$.

Let us call

$Cu(U, V, W, X, k)$ the assertion: "Z is linearly equivalent to 0 over the field k," and

$Sq(U, V, W, X, k)$ the assertion "Y is linearly equivalent to 0 over the field $k(u_0, u_1, v_0, v_1)$."

We shall prove the following theorems.

THEOREM 1. *The assertion $Sq(U, V, W, X, k)$ implies $Cu(U, V, W, X, k)$, and conversely, the assertion $Cu(U, V, W, X, k)$ implies $Sq(U, V, W, X, k(w'))$ for every simple point w' of W such that $U \times V \times w'$ is not contained in X.*

THEOREM 2. *There exists a field K such that both assertions are true relative to K. If W is complete and non-singular in codimension* 1, *then* $Sq(U, V, W, X, k)$ *is true.*

The assertion that Z is linearly equivalent to 0 will be called the *theorem of the cube*, while the assertion that Y is linearly equivalent to 0 will be called the *theorem of the square*. In Theorem 2, once we shall have proved the theorem of the square relative to some field K containing k, we can apply Corollary 1 of the last theorem in *Foundations* (or also IAG-VI_5 Cor. 2) to conclude that Y is in fact the divisor of a function defined over k, i.e., that Y is linearly equivalent to 0 over k, provided that W is complete and non-singular in codimension 1. The second part of Theorem 2 is therefore an immediate consequence of the first in this case. In the applications, it is the only one which we shall consider.

Let us now proceed to the proof of the two theorems.

Proof of Theorem 1: Let $Q = (u_0, u_1, v_0, v_1)$ be a generic point of $U^{(2)} \times V^{(2)}$ over k. We have

$$Z \cdot (Q \times W^{(2)}) = Q \times [(Y \times W) - (W \times Y)]$$

and consequently, after taking the intersection with

$$U^{(2)} \times V^{(2)} \times W \times w'$$

we find

$$Z \cdot (Q \times W \times w') = Q \times Y \times w'$$

for every simple point w' of W such that $U \times V \times w' \not\subset X$.

Let us assume assertion Cu and let F be a function defined over k on $U^{(2)} \times V^{(2)} \times W^{(2)}$, such that $Z = (F)$. According to Proposition 3 of § 1, we see that $Q \times Y \times w'$ is the divisor of the function induced by F on $Q \times W \times w'$, thus proving assertion Sq.

Conversely, assume that we can write $Y = (f)$ for some function f on W, defined over $k(Q)$. Let w be a generic point of W over $k(Q)$. There exists a function f^* on $U^{(2)} \times V^{(2)} \times W^{(2)}$ defined by

$$f^*(Q, w) = f(w),$$

and we have $Q \times (f) = (f^*) \cdot (Q \times W)$ by F-VIII_2 Th. 1, Cor. 3. There exists a function F on the big product $U^{(2)} \times V^{(2)} \times W^{(2)}$ defined over k such that

$$F(Q, w_0, w_1) = \frac{f^*(Q, w_0)}{f^*(Q, w_1)},$$

and one sees immediately from the definitions that (F) and Z have the same intersection with $Q \times W \times W$, and hence differ by a degenerate component of type $T \times W \times W$, where T is a divisor on $U^{(2)} \times V^{(2)}$ (Proposition 6 of the Appendix, § 2). On the other hand, the function induced by F on the diagonal $U^{(2)} \times V^{(2)} \times \Delta_W$ is equal to the constant 1, and the intersection of (F) with this variety is therefore equal to 0, by Proposition 3 of § 1. The same thing holds for the intersection of Z, because the terms coming from X_{ij0} and X_{ij1} will occur with opposite signs, and will cancel each other. This shows that our degenerate component can be none other than 0, and concludes the proof of Theorem 1.

Proof of Theorem 2: Suppose first that W is a complete non-singular curve. We are going to see that $Sq(U, V, W, X, k)$ follows immediately from the properties of the Jacobian. Let $f : W \to J$ be a canonical map of W into its Jacobian J. If (u, v) are independent generic points of U, V, we have a rational map given by

$$(u, v) \to S_f(X(u, v))$$

of $U \times V$ into J. Theorem 3 of Chapter II, § 1 allows us to decompose this map into a sum of mappings $f_1 : U \to J$ and $f_2 : V \to J$, and one sees immediately that $S_f(Y) = 0$. From Abel's theorem, we conclude that Y is linearly equivalent to 0 on W. Since W is complete and non-singular, Y is in fact linearly equivalent to 0 over $k(u_0, u_1, v_0, v_1)$.

Observe that if we had the Picard variety, the same argument would give us our theorem for a variety of arbitrary dimension. Conversely, it is precisely the possibility of giving a direct proof of the theorem of the square which will allow us to construct the Picard variety.

It is now tempting to proceed by induction on W, say by taking a generic hyperplane section. However, we then need what is known as an equivalence criterion, and attempts to prove such a criterion directly lead into trouble. It is almost a miracle that by using the theorem of the cube we can conclude the proof in a few lines, in the following manner:

By Theorem 1, we know that $Cu(U, V, W, X, k)$ is true if W is a complete non-singular curve. But our assertion Cu is *symmetric in the three factors* U, V, W. Hence $Cu(U, V, W, X, k)$ is true if any one of the three varieties U, V, W is a complete non-singular curve. Theorem 1 shows that there exists a field K such that $Sq(U, V, W, X, K)$ is true if, for instance, U is a complete non-singular curve. But in assertion Sq, the variety U plays the role of a parameter variety, and since the divisor Y is obtained by intersection over a generic point (u_0, u_1, v_0, v_1), it follows that Sq is birational with respect to U. Hence there exists a field K such that $Sq(U, V, W, X, K)$ is true if U is an arbitrary curve. It is now going to be trivial to finish the proof by induction.

As we have just remarked, assertion Sq is birational in U. We may therefore assume that U is an affine variety. Let C_t be a generic curve on U over K. (For facts concerning generic curves, see IAG-VII$_6$.) Then C_t is defined over a purely transcendental extension $K(t)$ of K. Let (M_0, M_1, N_0, N_1) be a generic point of $C_t^{(2)} \times V^{(2)}$ over $K(t)$. Then M_0, M_1 are independent generic points of U over K, and hence (M_0, M_1, N_0, N_1) is a generic point of $U^{(2)} \times V^{(2)}$ over K. According to Lemma 1, there exists a divisor X' on $C_t \times V \times W$ such that

$$X'(M_i, N_j) = X(M_i, N_j).$$

This reduces the proof of assertion Sq to the case of curves, and taking into account the results obtained above, finishes our proof of Theorem 2.

For the applications, the generic theorem which we have just proved will not suffice, and we complete it by giving the analogous result for special points.

THEOREM 3. *Let X be a divisor on a product $U \times V \times W$, let a_0, a_1 be two points of U, and b_0, b_1 two points of V such that $X(a_i, b_j)$ is defined for $i, j = 0, 1$. Then*

$$\sum (-1)^{i+j} X(a_i, b_j) \sim 0$$

on W.

Proof: The proof consists essentially in repeating the arguments used in the first half of Theorem 1, starting with the theorem of the cube. Let X_{ijk} be defined as before on the product $U_0 \times U_1 \times V_0 \times V_1 \times W_0 \times W_1$. The intersection

$$\sum (-1)^{i+j+k} X_{ijk} \cdot (Q \times W \times W)$$

is defined, Q now denoting the point (a_0, a_1, b_0, b_1). If as before we put

$$Z = \sum (-1)^{i+j+k} X_{ijk}$$

but let now

$$Y = \sum (-1)^{i+j} X(a_i, b_j),$$

then we find

$$Z \cdot (Q \times W \times W) = Q \times [(Y \times W) - (W \times Y)].$$

Let w be a generic point of W, and take the intersection with $U_0 \times U_1 \times V_0 \times V_1 \times w \times W$. We get $Q \times w \times Y$, and since Z is linearly equivalent to 0, it follows from the corollary to Proposition 3 of § 1 that $Y \sim 0$.

§ 3. *The theorem of the square for groups*

We shall use Theorem 3 mainly when we have an algebraic family of cycles whose parameter variety is a group variety (in fact an abelian variety).

Let Z be a cycle on a product $G \times W$ of a group variety with an arbitrary variety. Let a be a point of G. We denote by Z_a the cycle obtained by transforming Z by the biholomorphic correspondence of $G \times W$ on itself which sends each point (u, w) on (au, w). We may say that Z_a is the translation of Z by a on $G \times W$, even though it is only a translation on G.

As we proved in Chapter I, § 2 that the translation can be obtained by an intersection, we point out here that our transformed cycle Z_a can also be obtained by an intersection. Suppose first that Z is a variety, and k is a field of definition for G, W, and Z. If (v, w) is a generic point of Z over k, and u a generic point of G over

k independent of (v, w) then we let $\bar{Z}$ be the locus of (u, uv, w) over k. We extend the operation $Z \to \bar{Z}$ to cycles by linearity. According to F-VII$_6$ Th. 12 we get

$$\bar{Z} \cdot (u \times G \times W) = u \times Z_u.$$

Let a be an arbitrary point of G, and let us take on $G \times G \times W$ the translation by (au^{-1}, au^{-1}), i.e., the transformation which to each point (x, y, z) assigns the point $(au^{-1}x, au^{-1}y, z)$. This transformation sends $u \times G \times W$ on $a \times G \times W$, it leaves $\bar{Z}$ invariant, and sends $u \times Z_u$ on $a \times Z_a$. It follows that for every point $a \in G$, the intersection $\bar{Z} \cdot (a \times G \times W)$ is defined, and we have

$$\bar{Z} \cdot (a \times G \times W) = a \times Z_a.$$

We can then prove the following application of Theorem 3.

THEOREM 4. *Let G be a group variety, W an arbitrary variety, and X a divisor on the product $G \times W$. Let a_1, a_2, a_3, a_4 be four points of G such that $X(a_i)$ is defined $(i = 1, \ldots, 4)$ and such that $a_1 a_3^{-1} = a_2 a_4^{-1}$. Then the divisor*

$$X(a_1) - X(a_2) - X(a_3) + X(a_4)$$

is linearly equivalent to 0 on W.

Proof: Let T be the divisor

$$a_1^{-1}X - a_2^{-1}X - a_3^{-1}X + a_4^{-1}X$$

where we denote by $a^{-1}X$ the divisor $X_{a^{-1}}$. This is more appropriate since we make a left translation. Then the intersection $T \cdot (e \times W)$ is defined, and we have

$$T(e) = X(a_1) - X(a_2) - X(a_3) + X(a_4).$$

In view of the corollary to Proposition 3 of § 1, it will suffice to prove that $T \sim 0$. A translation by a point u of G sending (v, w) onto (uv, w) gives a biholomorphic transformation of $G \times W$ onto itself. To show that $T \sim 0$ it will therefore suffice to show that $uT \sim 0$, for some generic point u of G over a field of definition k for G, W over which X and all the points a_i are rational. Put $u_i = ua_i^{-1}$. The notation being the same as in the remarks at the beginning of this section, we have

$$uT = \bar{X}(u_1) - \bar{X}(u_2) - \bar{X}(u_3) + \bar{X}(u_4).$$

Let v be a generic point of G independent of u over k. Taking into account the fact that $u_1^{-1}u_3 = a_1u^{-1}ua_3^{-1} = a_1a_3^{-1}$ is rational over k, we see that the four points

$$(u_1v, v^{-1}), \quad (u_2v, v^{-1}), \quad (u_1v, v^{-1}u_1^{-1}u_3), \quad (u_2v, v^{-1}u_1^{-1}u_3)$$

are generic points of $G \times G$ over k (but of course are not independent). According to F-VII$_6$ Th. 12 there exists a divisor Y on $G \times G \times W$ such that

$$\begin{aligned}\bar{X}(u_1) &= Y(u_1v, v^{-1}),\\ \bar{X}(u_2) &= Y(u_2v, v^{-1}),\\ \bar{X}(u_3) &= Y(u_1v, v^{-1}u_1^{-1}u_3),\\ \bar{X}(u_4) &= Y(u_2v, v^{-1}u_1^{-1}u_3).\end{aligned}$$

Indeed, there exists an isomorphism of $k(u_1v, v^{-1})$ on each one of the fields obtained by adjoining the other three generic points of $G \times G$. The equalities involving $\bar{X}(u_2)$ and $\bar{X}(u_3)$ are then immediate, once we have picked a divisor Y such that $\bar{X}(u_1) = Y(u_1v, v^{-1})$ by F-VII$_6$ Th. 12. The fourth equality concerning $\bar{X}(u_4)$ now comes from the hypothesis which gives us $u_2u_1^{-1}u_3 = u_4$. Applying Theorem 3, we conclude the proof of our theorem.

COROLLARY 1. *Let G be a commutative group variety, W any variety, and X a divisor on the product $G \times W$. Let a_i be points of G $(i = 1, \ldots, n)$ and m_i integers such that $\sum m_i = 0$ and $\sum m_i \cdot a_i = 0$ (the law of composition being written additively). Denote as before by X_a the transform of X by the biholomorphic map of $G \times W$ sending a point (u, w) on $(u + a, w)$. Then,*

$$\sum m_i X_{a_i} \sim 0$$

on $G \times W$. The mapping $a \to \mathrm{Cl}\,(X_a - X)$ is a homomorphism of G into Pic $(G \times W)$.

Proof: Let $\bar{X}$ be the divisor on $G \times G \times W$ defined as above. We have $X_{a_i} = \bar{X}(a_i)$, and in particular, $X = \bar{X}(0)$. The hypothesis $\sum m_i = 0$ gives us the equality

$$\sum m_i X_{a_i} = \sum m_i[\bar{X}(a_i) - \bar{X}(0)]$$

and from Theorem 4 we see immediately that the map $a \to \mathrm{Cl}\,(X_a - X)$ is a homomorphism of G into Pic $(G \times W)$. Since we have $\sum m_i \cdot a_i = 0$ by hypothesis, this gives us $\sum m_i X_{a_i} \sim 0$.

COROLLARY 2. *Let G be a commutative group variety, W an arbitrary variety, and X a divisor on the product $G \times W$. Let a_i $(i = 1, \ldots, n)$ be points of G such that $X(a_i)$ is defined for each i. Let m_i be integers such that $\sum m_i = 0$ and $\sum m_i \cdot a_i = 0$. Then*

$$\sum m_i X(a_i) \sim 0$$

on W.

Proof: Let T be the divisor $\sum m_i X_{-a_i}$. Then $T \cdot (0 \times W)$ is defined, and $T(0)$ is equal to $\sum m_i X(a_i)$. Corollary 2 shows that $T \sim 0$ on $G \times W$. We can now use Proposition 3 of § 1 to prove what we want.

COROLLARY 3. *Let G be a commutative group variety, and W an arbitrary variety. Let X be a divisor on $G \times W$. Let a be a point of G, and k a field of definition for G, W over which X is rational. Let u be a generic point of G over $k(a)$. Then the mapping*

$$a \to \mathrm{Cl}\,[X(a + u) - X(u)]$$

induces a homomorphism of G into Pic (W).

Proof: It is an immediate consequence of Corollary 2.

REMARK. In Corollary 3, we have surmounted the technical difficulty that $X(0)$ may not be defined by making a generic translation (Chapter I, § 2, Proposition 7) and using Corollary 2. It gives us from now on complete technical freedom in working with algebraic families parametrized by commutative group varieties, provided we are interested only in the linear equivalence classes of divisors.

COROLLARY 4. *Let G be a commutative group variety, and X a divisor on G. Then the mapping*

$$\varphi_X : a \to \mathrm{Cl}\,(X_a - X)$$

is a homomorphism of G into Pic (G).

Proof: This is a consequence of Corollary 1 in the case when W is reduced to a point.

The homomorphism in Corollary 4 is of great importance, and will play a crucial role in the sequel. If X, Y are two divisors on G, then we obviously have

$$\varphi_{X+Y} = \varphi_X + \varphi_Y$$

and hence those divisors which are such that $\varphi_X = 0$ form a subgroup of the group of divisors on G. If $\varphi_X = 0$, we say that X is *squarely equivalent to* 0 and write $X \equiv 0$. We shall study this new equivalence relation especially when the commutative group is an abelian variety, but for the moment we can already prove a partial result relating it to algebraic equivalence.

PROPOSITION 4. *Let G be a commutative group variety, and let X be a divisor on G algebraically equivalent to* 0. *Then $\varphi_X = 0$, i.e.,* $X \equiv 0$.

Proof: According to Proposition 1 of § 1, we can find an abelian variety J and a divisor Z on $J \times G$ such that

$$X = Z(x) - Z(y)$$

with two suitable points x, y of J. Let T be the divisor on $J \times G$ given by

$$T = Z_{(-x,u)} - Z_{(-y,u)} - Z_{(-x,0)} + Z_{(-y,0)}$$

where $Z_{(w,u)}$ denotes, as usual, the transform of Z by the translation (w, u) on $J \times G$, and u is an arbitrary point of G. If we take $w = -x$ and u arbitrary, then

$$Z \cdot (x \times G) = x \times Z\,(x)$$

is transformed by the translation (w, u) into

$$Z_{(-x,u)} \cdot (0 \times G) = 0 \times Z(x)_u.$$

We have a similar formula if we take $w = -y$. This shows that

$$X_u - X = \mathrm{pr}_G\,[T \cdot (0 \times G)\,].$$

But according to Theorem 4, Corollary 1 we have $T \sim 0$ on $J \times G$, and hence $X_u - X \sim 0$ on G by Proposition 3 of § 1. Since u is arbitrary, this proves our proposition.

In the next chapter we shall prove that, conversely, if $X \equiv 0$ then X is a torsion divisor (on an abelian variety).

§ 4. *The kernel in the theorem of the square*

In the preceding section, we have seen that if X is a divisor on a product $G \times W$ of a commutative group variety with an arbitrary variety, then the mapping

$$a \to \mathrm{Cl}\,[X(a + u) - X(u)]$$

gives a homomorphism of G into Pic (W), the expression on the right-hand side being independent of the auxiliary generic point u selected to make the intersections defined. We shall prove in this section that, when W is complete and non-singular in codimension 1 and X is positive, the kernel is an algebraic subgroup of G. For this purpose, we need a method which allows us to take the quotient rationally, and the following discussion will give us this method.

Let V^r be a variety, complete and non-singular in codimension 1, defined over k. Let P be a simple point of V, rational over k. Let Ω be the universal domain, and let $t_1, \ldots, t_r$ be functions on V, defined over k, which are local uniformizing parameters of P on V. Denote by $\mathfrak{o}$ the local ring of P over Ω and by $\mathfrak{m}$ its maximal ideal. Then $t_1, \ldots, t_r$ are generators of $\mathfrak{m}/\mathfrak{m}^2$ over $\mathfrak{o}/\mathfrak{m}$, and every function $f \in \mathfrak{o}$ can be written as a power series in monomials of the t's, with coefficients in Ω. Denote by $\{M_j(t)\}$ all the monomials $t_1^{\nu_1} \ldots t_r^{\nu_r}$ in any given fixed order $j = 0, 1, 2, \ldots$ subject to the sole condition that $\deg M_j \leqq \deg M_{j+1}$. We can then write for $f \in \mathfrak{o}$,

$$f = \sum_{j=0}^{\infty} c_j M_j(t), \qquad c_j \in \Omega.$$

If f is defined over a field $K \supset k$, then the c_j are in K. The subring of $\mathfrak{o}$ consisting of those functions f defined over K will be denoted by $\mathfrak{o}^K$ and its maximal ideal by $\mathfrak{m}^K$.

The subset of functions of $\mathfrak{o}$ whose power series are such that

$$c_0 = c_1 = \ldots = c_N = 0$$

is obviously an ideal I_N of $\mathfrak{o}$, and these ideals $\{I_N\}$ form a decreasing sequence of ideals whose intersection is equal to 0 by Krull's theorem. We denote by $I_N{}^K$ the intersection of I_N with $\mathfrak{o}^K$, so that $I_N{}^K$ is the set of functions of $\mathfrak{o}$, defined over K, whose first $N + 1$ coefficients in the power series development are equal to 0. We have $\mathfrak{o} = I_{-1}$.

When such an ordered sequence of monomials $\{M_j(t)\}$, or equivalently a decreasing sequence of ideals $\{I_N\}$, is given, we shall say that we have defined a *system of linear conditions at P over k*. If $f \in \mathfrak{o}$ is in I_N we say that f *satisfies the linear conditions I_N at P*.

Let $\mathscr{L}$ be a linear system. (For the elementary theory and definitions concerning linear systems, see IAG-VI$_3$.) Let X be a divisor in $\mathscr{L}$. We can find a function f and a divisor Y on V such that $X = (f) + Y$ and $P \notin \operatorname{supp}(Y)$ (Appendix, § 2). This implies that f has no poles going through P, and hence that f is in $\mathfrak{o}$ (IAG-VI$_1$ Prop. 3). If we can write $X = (g) + Z$ with some other function g and a divisor Z such that $P \notin \operatorname{supp}(Z)$, then one sees that fg^{-1} is a unit in $\mathfrak{o}$ because it has neither a pole nor a zero going through P. Hence $f \in I_N$ if and only if $g \in I_N$. We shall say that X *satisfies the linear conditions I_N at P* if $f \in I_N$. We have just seen that this property does not depend on the auxiliary function f.

PROPOSITION 5. *Let V be complete, non-singular in codimension 1. Let $\mathscr{L}$ be a linear system on V, and assume given linear conditions $\{I_N\}$ at a simple point P of V. Let $\mathscr{L}_N$ be the subset of $\mathscr{L}$ consisting of all divisors $X \in \mathscr{L}$ satisfying the conditions I_N. Put $\mathscr{L} = \mathscr{L}_{-1}$. Then $\mathscr{L}_N$ is a linear system (as long as it is not empty), and it is empty for all sufficiently large N. Furthermore, we have*

$$\dim \mathscr{L}_{N+1} = \dim \mathscr{L}_N \quad \textit{or} \quad \dim \mathscr{L}_{N+1} = \dim \mathscr{L}_N - 1.$$

Proof: Recall first that if $\mathscr{L}$ is a linear system and L is a vector space of functions derived from $\mathscr{L}$, then $\dim \mathscr{L} = \dim L - 1$.

Now let X_0 be in $\mathscr{L}_N$. If X, X' are in $\mathscr{L}_N$ and if f, f' are two functions such that $(f) = X - X_0$ and $(f') = X' - X_0$, we must show that for all constants c, the divisor X_c such that $(f + cf') = X_c - X_0$ is in $\mathscr{L}_N$. Let g be a function and Y a divisor such that $X_0 = (g) + Y$, and $P \notin \operatorname{supp}(Y)$. Then we have

$$(fg) = X - Y, \qquad (f'g) = X' - Y.$$

This shows that fg and $f'g$ are in I_N, and hence that $(f + cf')g$ is in I_N. We obviously have

$$(f + cf')g = X_c - Y$$

and, by definition, we see that X_c satisfies the conditions I_N. This proves that $\mathscr{L}_N$ is a linear system.

A divisor in a linear system cannot satisfy the linear conditions I_N for all N at the point P. Hence as N increases, the dimension of the systems $\mathscr{L}_N$ must decrease, so that eventually $\mathscr{L}_N$ becomes empty.

We still have to prove that if N increases by 1 then the dimension of $\mathscr{L}_N$ can decrease by at most 1. This is essentially clear because we impose one linear condition at a time. More precisely, let S_N be the vector space (over the constants) of those power series for which $c_0 = c_1 = \ldots = c_N = 0$. It is of course infinite dimensional, but obviously the factor space S_N/S_{N+1} has dimension 1. On the other hand, let X_0 be in $\mathscr{L}_N$, and write $X_0 = (f_0) + Y_0$ with $P \notin \operatorname{supp}(Y_0)$. Each X in $\mathscr{L}_N$ determines a function f up to a multiplicative constant in the vector space L_N, derived from X_0, namely

$$(f) = X - X_0 = X - (f_0) - Y_0,$$

and conversely. We can thus write

$$(f_0 f) = X - Y_0$$

and $f_0 L_N$ is contained in $\mathfrak{o}$, and in fact in I_N and S_N. A divisor $X \in \mathscr{L}_N$ is in $\mathscr{L}_{N+1}$ if and only if the associated function $f_0 f$ is in S_{N+1}. Since S_N/S_{N+1} has dimension 1, it follows immediately that the dimension of $\mathscr{L}_{N+1}$ can be at most one less than that of $\mathscr{L}_N$.

They are equal of course if and only if it happened that $f_0 L_N$ is already contained in I_{N+1}.

PROPOSITION 6. *Let V be complete and non-singular in codimension 1. Let $\mathscr{L}$ be a complete linear system on V, and let $\{I_N\}$ be a system of linear conditions at a simple point P of V. Then:*

(i) *There exists an integer $N \geqq -1$ uniquely determined by the following property. At least one divisor X of $\mathscr{L}$ satisfies conditions I_N, and no divisor of $\mathscr{L}$ satisfies conditions I_{N+1}. For this integer N, there exists one and only one divisor X of $\mathscr{L}$ satisfying conditions I_N.*

(ii) *If V is defined over k, P is rational over k, $\{I_N\}$ is defined over k, and if there exists a divisor D on V, rational over a field $K \supset k$, and linearly equivalent to the divisors of $\mathscr{L}$, then the unique divisor X of* (i) *is rational over K.*

Proof: Our assertion (i) is an immediate consequence of Proposition 5. Let us prove (ii). Using the fact that a divisor is locally linearly equivalent to 0 at a simple point, we may assume that $P \notin \operatorname{supp}(D)$. Let $f_0, \ldots, f_n$ be a basis of the vector space $L(D)$ (consisting of those functions f such that $(f) \geqq -D$), such that each function f_i is defined over K. Such a basis can be found in view of our hypotheses on V and $\mathscr{L}$. Each function $f \in L(D)$ can be written

$$f = f_\lambda = \sum_{i=0}^{n} \lambda_i f_i, \qquad \lambda_i \in \Omega.$$

Since none of the f_i has a pole through P, each f_i is in $\mathfrak{o}$ and has an expression as a power series,

$$f_i = \sum_{j=0}^{\infty} c_{ij} M_j(t), \qquad c_{ij} \in K,$$

and consequently

$$f_\lambda = \sum_{j=0}^{\infty} \left(\sum_{i=1}^{n} \lambda_i c_{ij} \right) M_j(t).$$

The infinite matrix $\| c_{ij} \|$ has rank $n + 1$, because the f_i are linearly independent over K. Let s be the smallest integer such that $\| c_{ij} \|$ ($i = 0, \ldots, n$ and $j = 0, \ldots, s$) has rank $n + 1$. Then the equations

$$\sum_{i=0}^{n} \lambda_i c_{ij} = 0, \qquad j = 0, \ldots, s,$$

determine the λ_i uniquely in K (up to a multiplicative constant). The corresponding function f_λ has a divisor

$$(f_\lambda) = X - D$$

with $X > 0$. Since f_λ is defined over K, and D is rational over K, it follows that X is rational over K. It is clear from our construction that X is the divisor whose existence and uniqueness have been stated in (i). This concludes the proof.

THEOREM 5. *Let V be a complete variety, non-singular in codimension 1. Let G be a commutative group variety, and let D be a positive divisor on $G \times V$. Let k be a field of definition for G and V over which D is rational, and assume that V has a simple point P rational over k. Then there exists a commutative group variety B, a surjective homomorphism $\varphi : G \to B$ defined over k, and a positive divisor E on $B \times V$ rational over k satisfying the following conditions:*

(i) *If z is a generic point of B over k, then $k(z)$ is the smallest field of rationality of the divisor $E(z)$ (containing k).*

(ii) *If there exists a divisor X on V rational over a field $K \supset k$, such that $E(z) \sim X$, then z is rational over K.*

(iii) *If z, w are two generic points of B over k (not necessarily independent) then $E(z) \sim E(w)$ if and only if $z = w$.*

(iv) *If u is a generic point of G over k, then*

$$D(u) \sim E(\varphi(u)).$$

Before proving the theorem, let us first indicate some important consequences.

COROLLARY 1. *The kernel of $\varphi : G \to B$ is the same as the kernel of the homomorphism $v \to \mathrm{Cl}\,[D(v+u) - D(u)]$.*

Proof: The condition $\varphi(a) = 0$ is equivalent to the condition $\varphi(a+u) = \varphi(u)$ for u generic over $k(a)$, which in turn is equivalent to $D(a+u) \sim D(u)$ by (iii) and (iv).

COROLLARY 2. *Let X be a positive divisor on an abelian variety A. Then the kernel of the map $\varphi_X : u \to \mathrm{Cl}\,(X_u - X)$ is an algebraic subgroup of A.*

Proof: If we put $D = s_2^{-1}(X)$, we know from Chapter I, § 2 that $D(u) = X_{-u} - X$. Since the abelian variety has a simple rational point, namely the origin, we are brought back to the preceding corollary.

Proof of Theorem 5: We suppose given a system of linear conditions at P. As we have seen, for each complete linear system $\mathscr{L}$ on V, these conditions determine a unique positive divisor of $\mathscr{L}$, rational over the intersection of all fields of rationality of divisors linearly equivalent to those of $\mathscr{L}$. We shall denote by $\overline{D(u)}$ the unique divisor in the complete linear system $\mathscr{L}(D(u))$ determined by $D(u)$, for a generic point u of G over k. Then in particular, $\overline{D(u)}$ is rational over $k(u)$, and its smallest field of rationality containing k can be written $k(z)$ where z is the generic point of some variety W, defined over k, model of the function field $k(z)$ over k. We have a rational map $f : G \to W$ such that $f(u) = z$. We are going to define a normal law of composition on W. This will allow us to take for W a group variety, and we shall see that there is a surjective homomorphism of G onto W which is the desired one.

If u, v are two independent generic points of G over k, then

$$D(u + v) \sim D(u) + D(v) + D(w + t) - D(w) - D(t)$$

for any independent generic points w, t of u, v. (The auxiliary divisor $D(w + t) - D(w) - D(t)$ has been used only because $D(0)$ may not be defined.) The above relation is merely an application of the theorem of the square, Corollary 2 of Theorem 4, § 3. We can replace $D(u)$ and $D(v)$ by $\overline{D(u)}$ and $\overline{D(v)}$ in the above linear equivalence. This shows that $\overline{D(u + v)}$ is rational over $k(f(u), f(v), w, t)$ and hence over $k(f(u), f(v))$ since w, t can be selected arbitrarily. Therefore, $f(u + v)$ is rational over this field.

Let now x, y be two independent generic points of W over k. We can select two independent generic points u, v of G such that $x = f(u)$ and $y = f(v)$. I contend that the point $f(u + v) = z$ does not depend on the choice of u, v subject to the above conditions.

Once this has been shown, it will be clear that we have a law of composition $\psi : W \times W \to W$ defined by $\psi(x, y) = z$ satisfying all the conditions of a normal law of composition (Chapter I, § 1). The proof of our contention is based on the following lemma.

LEMMA 3. *Let u, v be two generic points of G over k (not necessarily independent). Then $\overline{D(u)} = \overline{D(v)}$ if and only if $f(u) = f(v)$.*

Proof: Suppose that $\overline{D(u)} = \overline{D(v)}$. Then $\mathscr{L}(D(u)) = \mathscr{L}(D(v))$ because these two complete linear systems have a divisor in common. Let $\sigma : k(u) \to k(v)$ be the isomorphism mapping u on v, and extend σ to an automorphism of the universal domain. Then σ transforms $\mathscr{L}(D(u))$ into itself, and hence maps the uniquely determined divisor of $\mathscr{L}(D(u))$ into itself, because σ leaves the rational point fixed, and the linear conditions are determined rationally over k. That is to say, the divisor $\overline{D(u)} = \overline{D(v)}$ remains fixed under σ. Its smallest field of rationality $k(f(u))$ is therefore also fixed under σ (IAG-III$_6$ Th. 12) and since $f(u)^\sigma = f(u^\sigma) = f(v)$, we see that $f(u) = f(v)$. Conversely, if $f(u) = f(v)$, then σ leaves the smallest field of rationality of $\overline{D(u)}$ fixed, and hence leaves $\overline{D(u)}$ fixed. Since σ transforms $\overline{D(u)}$ into $\overline{D(v)}$, this shows that these two divisors are equal.

Returning to the proof that z does not depend on the choice of u, v suppose that we have $x = f(u_1)$ and $y = f(v_1)$ with u_1, v_1 generic independent over k. There exists an isomorphism

$$\tau : k(u, v) \to k(u_1, v_1)$$

mapping u on u_1 and v on v_1. Our lemma yields $\overline{D(u)} = \overline{D(u_1)}$ and $\overline{D(v)} = \overline{D(v_1)}$. The smallest field of rationality of each one of these divisors, which are uniquely determined by our origin and our linear condition, is therefore fixed under τ. This implies that x and y are fixed under τ. But as we have seen, the point $z = f(u + v)$ is rational over $k(x, y)$. We have therefore

$$z = z^\tau = (f(u + v))^\tau = f(u^\tau + v^\tau) = f(u_1 + v_1).$$

This proves that z does not depend on the choice of points u, v.

Rewriting an earlier relation, we get

$$D(u) \sim D(u + v) - D(v) + D(w + t) - D(w) - D(t),$$

and we see in a similar manner as before that $f(u)$ is rational over $k(f(u+v), f(v))$. The associativity condition being trivially satisfied, we see from the definitions that our rational map gives us a normal law of composition on W, obviously commutative. We may therefore assume that W has been chosen as a commutative group variety B. The rational map $f: G \to B$ is then a generic homomorphism by construction, and is therefore a homomorphism (Chapter I, § 1). We observe that if G is complete, then B is also complete.

In addition, the divisor $\overline{D(u)}$ is rational over $k(f(u))$, and hence by F-VII$_6$ Th. 12 there exists a positive divisor E on $B \times V$ such that $\overline{D(u)} = E(f(u))$. Conditions (i) and (ii) are obviously satisfied. Condition (iii) is precisely what we have proved in the lemma, and (iv) has just been taken care of. This concludes the proof.

Historical Note:

This chapter is entirely due to Weil.

The useful lemmas concerning algebraic equivalence appear already in his attempt to construct the Picard variety through the equivalence criteria [89].

In his treatise [85] the theorem according to which the mapping $u \to \mathrm{Cl}\,(X_u - X)$ is a homomorphism on an abelian variety plays an essential role. Other special cases of the theorem of the cube are mixed together with the equivalence criteria [89], and it is only recently that he succeeded in extracting the general formulations of the theorems of the square and of the cube (unpublished) in the form in which they appear here, and in showing how they can be taken as basic principles, which can be proved directly from intersection theory and the properties of the Jacobian variety.

One should also note the elegant method used to construct rationally the kernel in the theorem of the square, by means of linear conditions at a simple point. The general procedure by which one constructs a quotient by taking a smallest field of rationality is of course quite similar to that used in getting transformation spaces and homogeneous spaces [66], [91].

CHAPTER IV

Divisor Classes on an Abelian Variety

In the last chapter we defined various equivalence relations, and we shall now determine the structure of the factor groups for these equivalence relations in the group of divisors of an abelian variety A. We have inclusions

$$D(A) \supset D_\tau(A) \supset D_a(A) \supset D_l(A).$$

The Picard group $D_a(A)/D_l(A)$ will be given the structure of an abelian variety called the Picard variety. We shall prove also that a divisor X on A is a torsion divisor if and only if $X \equiv 0$ (i.e., $X_u \sim X$ for all $u \in A$). The map

$$X \to \varphi_X$$

will induce therefore an isomorphism of the Néron-Severi group $D(A)/D_\tau(A)$ into the group of homomorphisms of A into its Picard variety. It will be shown in a later chapter that this group is finitely generated.

Our homomorphism $\varphi_X : u \to \mathrm{Cl}\,(X_u - X)$ will be the cornerstone around which all our arguments are built. A divisor X is said to be *non-degenerate* if the kernel of φ_X is finite. To begin with, we prove that a positive divisor X is non-degenerate if and only if some positive multiple mX is ample (i.e., its linear system gives a projective embedding of A). We shall prove the existence of positive non-degenerate divisors, and we shall also see that if X is such a divisor, then φ_X maps A *onto* Pic (A). Using the last section of Chapter III, we then get the Picard variety of A trivially.

In addition to all this, we have to deal with numerical equivalence. One says that a cycle Z is numerically equivalent to 0 if deg $(Z \cdot Y) = 0$ for every cycle Y of complementary dimension such that $Z \cdot Y$ is defined. The cycle classes for numerical equivalence form a ring $N_*(A)$ under the intersection product. If $\alpha : A \to A$ is an endomorphism of A, then we can represent α on $N_*(A)$ by the inverse image $Z \to \alpha^{-1}(Z)$. We shall prove that the representation on the divisors is quadratic in α. If $r = \dim A$, then raising the divisors to the rth power, we see immediately that the representation on points is of degree $2r$. In this way, we recover some classical statements concerning the points of finite order on A, for instance the fact that $\nu(n\delta) = n^{2r}$. We get the characteristic polynomial of α, namely $\nu(\alpha + n\delta)$ which is a polynomial of degree $2r$ in n with rational coefficients. Its penultimate coefficient is the trace $\text{tr}(\alpha)$, for which we shall give a canonical expression as an intersection number. We shall go more deeply into its properties in the next chapter.

It will be almost a triviality to show that a torsion divisor is numerically equivalent to 0. We shall wait until Chapter V to prove the converse. This will mean that on an abelian variety, the three equivalence relations: torsion, numerical, and of the square, amount to the same thing. As mentioned already, they are also identical with algebraic equivalence, but we shall prove it only up to a power of the characteristic (Chapter VII).

One more remark concerning the quadraticity of the above representation on divisors. Over the complex numbers, this can be seen immediately for a trivial topological reason: The homology ring is generated by the topological 1-cycles, and the direct representation of α on them is linear in α. It suffices to observe that the algebraic 1-cycles correspond to a subset of the topological 2-cycles.

The corollaries to Theorem 4 of Chapter III, § 3 will be used constantly, and for convenience, we refer to any one of them as the *theorem of the square.*

§ 1. *Applications of the theorem of the square to abelian varieties*

In case the commutative group variety of Chapter III, § 4 is complete, i.e., is an abelian variety, one can obtain more precise results concerning the mapping $\varphi_X : a \to \mathrm{Cl}\,(X_a - X)$.

Let us first recall some general facts concerning the projective embedding of a variety by means of a complete linear system. Let V be a variety, complete and non-singular in codimension 1. Let X be a divisor on V. As usual, $\mathscr{L}(X)$ denotes the set of all positive divisors linearly equivalent to X, and $L(X)$ the set of functions φ on V such that $(\varphi) \geqq -X$. It is a vector space over the constants, of finite dimension. If $\mathscr{L}(X)$ is not empty, it is called a complete linear system. If V is defined over k, and X is rational over k, then one can find a basis for $L(X)$ defined over k. We say that $\mathscr{L}(X)$ *separates points* of V if for any two distinct points P, Q of V there exists a divisor $Y \in \mathscr{L}(X)$ such that $P \in \mathrm{supp}\,(Y)$ but $Q \notin \mathrm{supp}\,(Y)$. If $\varphi_0, \ldots, \varphi_m$ is a basis of $L(X)$ and P is a generic point of V, then the map

$$P \to (\varphi_0(P), \ldots, \varphi_m(P))$$

defines a generically surjective rational map of V onto a projective variety. The complete linear system $\mathscr{L}$ is said to be *ample* if this rational map is birational and biholomorphic. The existence of the projective embedding of a normal variety V is equivalent with the existence of an ample linear system on V. (All the above facts concerning linear systems can be found in IAG-V_4.)

THEOREM 1. *Let A be an abelian variety. Then there exists a projective embedding of A. If X is a positive non-degenerate divisor on A, then there exists an integer $m > 0$ such that mX is ample.*

Proof: Let us first prove the projective embedding. It suffices in fact to show the existence of a complete linear system which separates points of A. Indeed, the existence of such a system gives a rational map of A onto a projective variety U, which is one-one. This mapping may be purely inseparable, but the normalization of U in the function field of A is projective, and is biholomorphic to A according to Zariski's Main Theorem. (For details, see again IAG-VI_4 Prop. 8.) Let us show the existence of the desired system. Let $W^{(1)}, \ldots, W^{(m)}$ be subvarieties of A of codimension 1 (prime divisors) going through the origin, and such that their intersection is reduced to the origin. The proof of the existence of such varieties is a local problem on affine varieties, whose solution is obvious. Put $X = 3 \sum W^{(i)}$. Then we contend that the linear system $\mathscr{L}(X)$ separates points of A. Indeed, the linear equivalence class of X contains the divisor

$$\sum_{i=1}^{m} (W^{(i)}_{u_i} + W^{(i)}_{v_i} + W^{(i)}_{-u_i-v_i})$$

no matter what points u_i, v_i of A we take, by the theorem of the square on groups. Let a, b be two distinct points of A. By hypothesis, one of the $W^{(i)}$ does not contain $b - a$. Say it is $W^{(1)}$. Take $u_1 = a$, and take v_1, u_i, v_i $(i \neq 1)$ generic independent over $k(a, b)$, for some suitably large field k. Then the above divisor goes through a but not through b, thus proving our assertion.

We are going to prove the second part of the theorem by arguments similar to the preceding one. Let X be a positive non-degenerate divisor, and write X as a formal sum of varieties,

$$X = \sum X^{(i)}.$$

We may of course have repetitions among the $X^{(i)}$. For each i, the set of points $a \in A$ such that $X^{(i)}_a = X^{(i)}$ is contained in the kernel of φ_X and is therefore finite. Put

$$Y = \sum (X^{(i)}_{u_i} + X^{(i)}_{v_i} + X^{(i)}_{-u_i-v_i})$$

with arbitrary points u_i, v_i of A. Then $Y \sim 3X$. Let S be the kernel of φ_X, and a, b two points of A such that $b - a$ is not in S. For one of the varieties $X^{(i)}$, say $X^{(1)}$, there exists a point $c \in X^{(1)}$ such that $c + (b - a)$ is not in $X^{(1)}$. Put $u_1 = a - c$, and take v_1, u_j, v_j $(j > 1)$ independent generic points of A over $k(a, b, c)$ for a suitably big field k. One then sees that the linear system $\mathscr{L}(3X)$ separates pairs of points (a, b) such that $b - a$ is not in S, and in particular that for the rational map induced by this system, there is a bounded number of points in the inverse image of a point on the projective variety. By general elementary theorems concerning linear systems (IAG-VI_4 Prop. 8) we can conclude that $\mathscr{L}(nX)$ is ample for n sufficiently large.

For the convenience of the reader, let us recall briefly the argument which shows that under the above condition $\mathscr{L}(nX)$ is ample for large n. It is again a question of normalization. Let $T : A \to V$ be the rational map determined by $\mathscr{L}(Y)$. The hypothesis that S is finite shows that T is of finite degree, that $\mathscr{L}(Y)$ has no base points, and that T is everywhere defined. The inverse image $T^{-1}(Q)$ of a point Q of V contains only a finite number of points not exceeding the number of points in S. Since A is normal, we conclude that it is the normalization of V in the function field $k(A)$ of A. Let $f_0 = 1, \ldots, f_m$ be a basis of $L(Y)$ defined over k, assuming that k is a field of definition for A over which X and Y are rational. Then $k[f_0, \ldots, f_m]$ is an affine coordinate ring for an affine representative V_0 of V, and T induces a rational map $T_0 : A \to V_0$. Let V_0^* be the normalization of V_0 in the function field $k(A)$. Since the f_i have a pole only at components of Y, every function of $k(A)$ integral over $k[f_0, \ldots, f_m]$ has poles only at the components of Y. Hence there exists an integer e_0 such that the rational map induced by $\mathscr{L}(e_0 Y)$ is birational and biholomorphic at every point P of A such that $T(P)$ has a representative on V_0. We can repeat this argument a finite number of times, for affine varieties V_i covering V. We obtain each time an integer e_i, and it suffices to take $e \geqq e_i$ for all i in order that $\mathscr{L}(eY)$ be ample.

Note that we have not yet proved the existence of positive nondegenerate divisors. This will be done in the next section.

If the abelian variety is defined over k, then as usual we can find a projective embedding defined over k. Indeed, the proof of Theorem 1 shows that the varieties $W^{(i)}$ can be taken over the algebraic closure of k, and hence over a finite algebraic extension k' of k. Let X be an ample divisor on A, rational over k'. Taking the sum of X with its conjugates over k, with a multiplicity equal to a sufficiently large power of the characteristic, we obtain a divisor Y rational over k. All the conjugates of X are also ample, and hence the linear system $\mathscr{L}(Y)$ separates points of A. Since we know that there exists a basis of $L(Y)$ defined over k, we thus obtain a bijective rational map of A onto a projective variety defined over k, whose k-normalization is biholomorphic to A over k.

The projective embedding of A will be used in an essential manner in our treatment of numerical questions on an abelian variety.

We are now going to study more deeply the formalism of the equivalence $\equiv$.

As in Chapter I, § 2, let $A_{(m)}$ be the product $A \times \ldots \times A$ taken m times, and let $s_m : A_{(m)} \to A$ be the sum. If $u_1, \ldots, u_m$ are independent generic points of A, then

$$s_m(u_1, \ldots, u_m) = u_1 + \ldots + u_m.$$

We denote by $p_i : A_{(m)} \to A$ the projection on the ith factor. If X is a divisor on A, then $p_i^{-1}(X)$ is equal to the divisor

$$A \times \ldots \times X \times \ldots \times A$$

with X instead of A in the ith factor.

THEOREM 2. *Let X be a divisor on A. Then $X \equiv 0$ if and only if*

$$s_m^{-1}(X) \sim \sum_{i=1}^{m} p_i^{-1}(X).$$

Proof: In order to simplify the notation, let us take the case $m = 2$. For u generic on A, we have

$$s_2^{-1}(X) \cdot (A \times u) = X_{-u} \times u$$

and consequently

$$[s_2^{-1}(X) - X \times A] \cdot (A \times u) = (X_{-u} - X) \times u.$$

If $X \equiv 0$, then the seesaw principle (Appendix § 2, Theorem 6) yields

$$s_2^{-1}(X) - X \times A \sim A \times Y$$

with a suitable divisor Y. By symmetry, we must have $Y \sim X$. The converse is obvious.

COROLLARY. *Let $\alpha : A \to B$ and $\beta : A \to B$ be two homomorphisms of an abelian variety A into an abelian variety B, and let X be a divisor on B, $X \equiv 0$. Let $\xi = \mathrm{Cl}\,(X)$. Let α^{-1}, β^{-1} be the induced* *homomorphisms of* $D(B)/D_l(B)$ *into* $D(A)/D_l(A)$. *Then*

$$(\alpha + \beta)^{-1}(\xi) = \alpha^{-1}(\xi) + \beta^{-1}(\xi).$$

In particular, if $A = B$, *and* n *is an integer, then*

$$(n\delta)^{-1}(\xi) = n \cdot \xi.$$

Proof: Let $\lambda = (\alpha, \beta)$ be the product mapping of A into $B \times B$, i.e., the map $\lambda(u) = (\alpha(u), \beta(u))$, and let s_2 be the sum on B. We have $\alpha + \beta = s_2\lambda$.

$$A \xrightarrow{\lambda} B \times B \xrightarrow{s_2} B$$

Let p_i $(i = 1, 2)$ be the projection of $B \times B$ on the corresponding factor. Then $p_i\lambda = \alpha$ or β according as $i = 1$ or 2. Let X_u be a generic translation of X. Then $X_u \sim X$ by hypothesis. We find

$$(\alpha + \beta)^{-1}(X_u) = \lambda^{-1}(s_2^{-1}(X_u)) \sim \lambda^{-1}(X_u \times B + B \times X_u)$$

because all these expressions are defined (Proposition 7 of Chapter I, § 2) and we can use Corollary 3 of Theorem 3, Appendix, § 1. On the other hand, we have

$$\alpha^{-1}(X_u) = \lambda^{-1}(p_1^{-1}(X_u)) = \lambda^{-1}(X_u \times B),$$
$$\beta^{-1}(X_u) = \lambda^{-1}(p_2^{-1}(X_u)) = \lambda^{-1}(B \times X_u).$$

This concludes the proof of our corollary.

Let us denote by $N'(A)$ the factor group of divisors on A by the subgroup of divisors $X \equiv 0$. We are now going to show that a homomorphism $\alpha : A \to B$ induces a homomorphism $\alpha^{-1} : N'(B) \to N'(A)$ which we shall also denote by α^{-1}. We shall always specify which equivalence relation we are working with, so as not to confuse the inverse homomorphisms induced by α on the linear equivalence classes and on $N'(B)$. We shall need a lemma.

LEMMA 1. *Let* $\lambda : A \to B$ *be a homomorphism,* X *a divisor on* B, *and* a *a point of* A *such that* $\lambda^{-1}(X_{\lambda a})$ *is defined. Suppose also that* $\lambda^{-1}(X)$ *is defined. Then*

$$(\lambda^{-1}(X))_a = \lambda^{-1}(X_{\lambda a}).$$

Proof: The translation $(a, \lambda a)$ on $A \times B$ leaves the graph of λ invariant. It transforms $\lambda^{-1}(X)$ on its translate by a, and transforms X on its translate by λa. This makes the lemma obvious.

Let now ξ be an element of $N'(B)$, and let X, Y be two divisors representing this class. I say that if $\lambda^{-1}(X)$ and $\lambda^{-1}(Y)$ are defined, then $\lambda^{-1}(X) \equiv \lambda^{-1}(Y)$. It suffices of course to show that if $X \equiv 0$ and if $\lambda^{-1}(X)$ is defined then $\lambda^{-1}(X) \equiv 0$. Let t be a generic point of B and u a generic point of A independent of t over a field of definition k for A, B, λ over which X is rational. Then $X \sim X_t$, and consequently $\lambda^{-1}(X) \sim \lambda^{-1}(X_t)$. It suffices to show that $\lambda^{-1}(X_t) \equiv 0$, taking into account Corollary 3 of Theorem 3, Appendix, § 1, and Proposition 4 of Chapter III, § 3. But $t + \lambda u$ is a generic point of B over k, and $\lambda^{-1}(X_{t+\lambda u})$ is defined. According to the lemma, we have

$$\lambda^{-1}(X_t)_u - \lambda^{-1}(X_t) = \lambda^{-1}(X_{t+\lambda u}) - \lambda^{-1}(X_t) = \lambda^{-1}(X_{t+\lambda u} - X).$$

Since we have $X_{t+\lambda u} - X \sim 0$ by hypothesis, we can apply the definition of the equivalence $\equiv$ to see that $\lambda^{-1}(X_t) \equiv 0$.

Proposition 1. *Let $\alpha : A \to B$ and $\beta : B \to C$ be two homomorphisms and let ξ be an element of $N'(C)$. Then*

$$\alpha^{-1}(\beta^{-1}(\xi)) = (\beta\alpha)^{-1}(\xi).$$

Proof: A generic translation allows us to find a representative divisor X of ξ such that all intersections with which we are concerned are defined. We can then apply Theorem 2 of the Appendix.

Proposition 2. *Let ξ be an element of $N'(A)$ and n an integer. Then*

$$(n\delta)^{-1}(\xi) = n^2 \cdot \xi.$$

More generally, let $\alpha_1, \ldots, \alpha_d : A \to B$ be homomorphisms of an abelian variety into another, and let $m_1, \ldots, m_d$ be integers. Let ξ be an element of $N'(B)$. Then we have

$$(m_1\alpha_1 + \ldots + m_d\alpha_d)^{-1}(\xi) = \tfrac{1}{2} \sum_{i,j} m_i m_j D_\xi(\alpha_i, \alpha_j)$$

where we denote by $D_\xi(\alpha, \beta)$ the element of $N'(A)$ given by the formula

$$D_\xi(\alpha, \beta) = (\alpha + \beta)^{-1}(\xi) - \alpha^{-1}(\xi) - \beta^{-1}(\xi).$$

(Remark: We have taken 1/2 in the above expression because we sum over all indices i, j, obtain each term twice, and thus must divide by 2.)

Proof: Let Y be a divisor representing ξ, and let k be a field of definition for A, B and all the α_i over which Y is rational. Let v be a generic point of B over k, and put $X = Y_v$. Then X is also a representative of ξ, and all the intersections which we are going to consider are going to be defined (Proposition 7 of Chapter I, § 2).

Put $\lambda = m_1\alpha_1 + \ldots + m_d\alpha_d$, and let u be a generic point of A. Lemma 1 shows that

$$\lambda^{-1}(X)_u - \lambda^{-1}(X) = \lambda^{-1}(X_{\lambda u} - X).$$

Since $X_{\lambda u} - X \equiv 0$ by the theorem of the square, we can use the corollary to Theorem 2 and a repeated application of the theorem of the square in order to see that the above cycle is linearly equivalent to

$$\sum_{i,j} m_i m_j \alpha_i^{-1}(X_{\alpha_j u} - X).$$

On the other hand, if we denote by $D_X(\alpha_i, \alpha_j)$ the divisor obtained by replacing ξ by X in the statement of our proposition, we get

$$D_X(\alpha_i, \alpha_j)_u - D_X(\alpha_i, \alpha_j) \sim \alpha_i^{-1}(X_{\alpha_j u} - X) + \alpha_j^{-1}(X_{\alpha_i u} - X)$$

always using Lemma 1, and the corollary to Theorem 2, together with the theorem of the square. If we multiply by $m_i m_j$, and if we take the sum for i, j we obtain twice the divisor occurring in the first part of the proof, up to linear equivalence. This concludes the proof of our proposition.

Let X be a divisor on A. As usual, we denote by X^- its transform by the automorphism of A sending u into $-u$. If $X \equiv Y$, then obviously $X^- \equiv Y^-$, and we denote by ξ^- the class of X^- in $N'(A)$. We have the following corollary.

COROLLARY. *For any element $\xi \in N'(A)$, we have $\xi = \xi^-$.*

Proof: This is immediate, because $\xi^- = (-\delta)^{-1}(\xi) = \xi$ by the quadraticity.

In order to complete the theory of the quadraticity in Proposition 2, it is preferable to wait until we have the formalism of the Picard variety.

§ 2. *The torsion group*

We are going to relate algebraic equivalence between divisors of A with the equivalence of the square, with which we have defined $N'(A)$.

THEOREM 3. *Let X be a positive divisor on A. Assume that there exists an abelian subvariety B of A such that if u is a generic point of B over some field of definition k of A, B over which X is rational, then $X_u \sim X$. Then $X_u = X$.*

Proof: The idea of the proof consists in representing B in the complete linear system $\mathscr{L}(X)$, and thus in the projective group. Such a representation can only be trivial. More precisely, let $f_0, \ldots, f_m$ be a basis of $L(X)$ such that all the f_i are defined over k. We may assume $f_0 = 1$. We can write $(f_i) = X_i - X$ with $X_i \in \mathscr{L}(X)$. If g is a function on A, we denote by g_u the function such that $g_u(v) = g(v - u)$. We then see immediately that if Y is the divisor of g, then $(g_u) = Y_u$. Hence we have

$$(f_{i,u}) = X_{i,u} - X_u.$$

By hypothesis, there exists a function $\varphi_{(u)}$ defined over $k(u)$ such that $X_u = X + (\varphi_{(u)})$ (Corollary 1 of the last theorem in *Foundations*). The divisor of the function $\varphi_{(u)} f_{i,u}$ is therefore equal to

$$X_u - X + X_{i,u} - X_u = X_{i,u} - X.$$

This shows that this function is in $L(X)$. Since it is defined over $k(u)$, we may write

$$\varphi_{(u)} f_{i,u} = \sum_j c_{ij}(u) f_j$$

the coefficients $c_{ij}(u)$ being constants, rational over $k(u)$. Denote by $\lambda(u)$ the element of the projective group determined by the matrix $\| c_{ij}(u) \|$. We are going to see that $u \to \lambda(u)$ is a representation of B in the projective group. If we apply a translation to the above linear combination by a generic point v of B independent from u over k, and if we multiply by $\varphi_{(v)}$, we get

$$\varphi_{(v)}\varphi_{(u),v}f_{i,u+v} = \sum_{j,s} c_{ij}(u)c_{js}(v)f_s.$$

In this formula, $\varphi_{(u),v}$ denoted the translation of $\varphi_{(u)}$ by v, and we have used the trivial fact that the successive translation of f_i by u and v is the same as by $u + v$. Since we have $(\varphi_{(u)}) = X_u - X$ by definition, the translation by v gives $(\varphi_{(u),v}) = X_{u+v} - X_v$. We also have $(\varphi_{(v)}) = X_v - X$. Taking the sum of these divisors and the product of the functions, we find

$$(\varphi_{(v)}\varphi_{(u),v}) = X_{u+v} - X = (\varphi_{(u+v)}).$$

Since two functions having the same divisor differ by a constant, we can find a constant $d(u, v)$ such that

$$d(u, v)\varphi_{(u+v)} = \varphi_{(v)}\varphi_{(u),v}.$$

We have

$$d(u, v)\varphi_{(u+v)}f_{i,u+v} = \sum_s c_{is}(u + v)d(u, v)f_s.$$

Together with the relation obtained above, we get

$$d(u, v)c_{is}(u + v) = \sum_j c_{ij}(u)c_{js}(v)$$

because the functions f_s are linearly independent over the constants This shows that the map $u \to \lambda(u)$ is a generic homomorphism of B into the projective group. We know that such a homomorphism is everywhere defined and that the image $\lambda(B)$ of B is a group subvariety of the projective group (Chapter I, § 1). Since B is complete, $\lambda(B)$ is also complete. This can be the case only if λ is trivial, i.e., maps B on the identity. It follows that we can write

$$\varphi_{(u)}f_{i,u} = c(u)\sum_j f_j$$

where $c(u)$ is a constant rational over $k(u)$. Taking $i = 0$, and using the fact that we choose $f_0 = 1$, we get $\varphi_{(u)} = c(u)$, or in other words, $\varphi_{(u)}$ is constant, and $X_u = X$, thus proving our theorem.

COROLLARY 1. *The kernel of the endomorphism $n\delta$ of an abelian variety is finite, and thus $\nu(n\delta) > 0$.*

Proof: The kernel is an algebraic subgroup, and its connected component is an abelian subvariety B of A annihilated by $n\delta$. Let X be a positive divisor on B and u a generic point of B. According to the theorem of the square, we have $n(X_u - X) \sim X_{nu} - X = 0$, and hence $(nX)_u = nX$, by the theorem. This is a relation between divisors, and hence $X_u = X$, which is absurd if $\dim B > 0$.

COROLLARY 2. *Let X be a divisor on A such that there exists an integer m for which mX is ample. Then X is non-degenerate.*

Proof: We have $\varphi_{mX}(u) = m \cdot \varphi_X(u) = \varphi_X(mu)$. Using Corollary 1 we see that it suffices to prove that mX is non-degenerate for some m. By hypothesis, mX is linearly equivalent to a hyperplane section of A in some projective embedding. From Proposition 4 of Chapter III, § 3, we know that if Y_1, Y_2 are two divisors which are linearly equivalent, then $\varphi_{Y_1} = \varphi_{Y_2}$. Hence it suffices to prove that a hyperplane section is non-degenerate. We know that if Y is positive, then the kernel of φ_Y is an algebraic group (Chapter III, § 4). By the theorem, the connected component of this group leaves Y invariant. If Y happens to go through the origin, then this connected component must be contained in Y. Now, in a projective embedding, all the hyperplane sections are linearly equivalent and are positive divisors. Let us consider only those which go through the origin. If B is an abelian subvariety of A leaving such a hyperplane section H invariant, then B is contained in H. This being the case for all hyperplane sections going through the origin, we conclude that B must be a point. This proves our corollary.

COROLLARY 3. *Let X be a positive divisor an A. In order that X be non-degenerate, it is necessary and sufficient that there exist an integer m such that mX is ample.*

Proof: This is merely a reformulation of Theorem 1 and the preceding corollary.

The next lemma will give us the connection between algebraic equivalence and the equivalence $\equiv$.

LEMMA 2. *Let $f : C \to A$ be a rational map of a complete non-*

singular curve into an abelian variety, which generates A. *Let* $r = \dim A$. *Let* W *be the divisor*

$$W = f(C) \oplus f(C) \oplus \ldots \oplus f(C)$$

the sum being taken $(r-1)$ *times. Let* X *be a divisor on* A, *and assume that* $X \equiv 0$. *Assume that the inverse image*

$$\mathfrak{a} = f^{-1}(X) = \sum m_j P_j$$

is defined. Let d *be the degree of the rational map sending* $C \times C \times \ldots \times C$ $(r$ *times) on* A; *and let* d_0 *be the degree of the rational map sending* $C \times \ldots \times C$ $(r-1$ *times) on* W. *Then*

$$dX \sim rd_0 \sum_j m_j W_{f(P_j)}.$$

Proof: Let us denote by $f_r : C \times \ldots \times C \to A \times \ldots \times A$ the product of f with itself r times, and let s_r be the sum on $A \times \ldots \times A$. Put $F = s_r f_r$. Then d is the degree of F. Theorem 5 of the Appendix shows that

$$FF^{-1}(X) = dX.$$

On the other hand, according to Theorem 2 of § 1 we get

$$F^{-1}(X) \sim \sum_{i=1}^{r} f_r^{-1} p_i^{-1}(X)$$

and the right-hand side of this expression is equal to

$$\sum_{i=1}^{r} \sum_j m_j (C \times \ldots \times P_j \times \ldots \times C).$$

We are going to apply F to this expression. Let G be the graph of F, and let P be a generic point of C. As in F-VII$_6$ Th. 12 let us take $G(P)$ on the product $C \times \ldots \times \hat{C} \times \ldots C \times A$ (omitting C in the ith place), that is to say, we take the projection on this variety of the intersection

$$G \cdot (C \times \ldots \times P \times \ldots \times C \times A).$$

It is a variety, whose set theoretic projection on A is $W_{f(P)}$. If we take its projection on A in the sense of intersection theory, i.e., if we take pr_A of the above cycle, we obviously get $d_0 W_{f(P)}$. If we now take a special point P_j, and if we use the compatibility of intersection and projection with specialization (for positive cycles), and the fact that $W_{f(P)}$ has the unique specialization $W_{f(P_j)}$ over $P \to P_j$ (Proposition 3 of Chapter I, § 2 with V equal to a point), we obtain the desired relation (using F-VIII$_2$ Th. 7)

$$dX \sim rd_0 \sum m_j W_{f(P_j)}.$$

In order to apply the preceding lemma, we must make two remarks. First, we can always find a generating curve, for instance by means of Lemma 6 of Chapter II, § 3 applied to the generic curve of a normal projective variety birationally equivalent to A, or even to A itself since we know now that A has a projective embedding.

Second, in Lemma 2 suppose that k is a field of definition for f over which X is rational. If at first $f^{-1}(X)$ is not defined, let us make a generic translation X_t of X by a generic point t of A over k. Then $X_t \sim X$ by hypothesis, and if X_v is another generic translation, then $X_t \sim X_v$. We know from Proposition 7 of Chapter I, § 2 that $f^{-1}(X_t)$ and $f^{-1}(X_v)$ are defined. By Corollary 3 of Theorem 3, Appendix, § 1, they are linearly equivalent. If we denote these divisors on C by $\mathfrak{a}_t$ and $\mathfrak{a}_v$ respectively, we must have $S(f(\mathfrak{a}_t)) = S(f(\mathfrak{a}_v))$ by Abel's theorem, together with the universal mapping property of the Jacobian. This shows that the point

$$u = rd_0 \sum_j m_j f(P_j) = rd_0 S(f(\mathfrak{a}))$$

does not depend on the generic translation which we might have had to make on X in order to have $f^{-1}(X)$ defined, because this point is rational over $k(t) \cap k(v) = k$.

We now apply the theorem of the square, and find

$$rd_0 \sum_j m_j (W_{f(P_j)} - W) \sim W_u - W \equiv 0.$$

Using the lemma, the hypothesis $X \equiv 0$, and Proposition 4 of Chapter III, § 3, we find

$$0 \equiv dX \equiv rd_0 (\sum m_j) W.$$

Hence there is an integral multiple of W which is $\equiv 0$. Theorem 3 shows that this can happen only if this multiple is equal to 0, or in other words $\sum m_j = 0$. From this we get immediately

$$dX \sim W_u - W.$$

We can summarize the preceding discussion in a theorem.

THEOREM 4. *There exists a positive divisor W on A, a field of definition k for A over which W is rational, and an integer d having the following property. For every divisor $X \equiv 0$ on A, there exists a point $u \in A$ such that $dX \sim W_u - W$. If X is rational over a field $K \supset k$, then we can choose u rational over K. If $f : C \to A$ is a rational map of a complete non-singular curve into A which generates A, then the divisor W on A obtained by taking the sum of $f(C)$ $r - 1$ times has the preceding property.*

Note that the canonical map of a curve into its Jacobian generates the Jacobian, and hence that the divisor Θ has the property stated in Theorem 4.

COROLLARY 1. *Let W be a positive divisor on A having the property stated in Theorem 4. Then the map*

$$\varphi_W : A \to \mathrm{Pic}\,(A)$$

is surjective, and W is non-degenerate.

Proof: Let X be a non-degenerate divisor rational over a field $k_1 \supset k$. The existence of such a divisor is guaranteed by Corollary 2 of Theorem 3. Let t be a generic point of A over k_1. Then there exists a point u of A such that

$$d(X_t - X) \sim W_u - W$$

and we can take u rational over $k_1(t)$ since $d(X_t - X)$ is rational over that field. We have $d \cdot \varphi_X(t) = \varphi_W(u)$. According to Theorem 4 of Chapter II, § 1 there exists an endomorphism α of A such that $\alpha t = u + c$ with a constant c, and we have $d \cdot \varphi_X(t) = \varphi_W(\alpha t) + \varphi_W(c)$. Since φ_X and $\varphi_W \alpha$ are homomorphisms, we must have $\varphi_W(c) = 0$, and we may therefore assume that $c = 0$. Since the kernel of φ_X is finite by hypothesis, it follows that the kernel of $\varphi_W \alpha$ is finite, and also the kernel of φ_W : W is non-degenerate.

Let us now prove the surjectivity of φ_W. Let Y be a divisor on

A, algebraically equivalent to 0. We can write $Y = Z(x) - Z(y)$, with a divisor Z on a product $J \times A$ of A with an abelian variety J, and with two points x, y of J which we may assume to be generic by Corollary 2 of Theorem 4, Chapter III, § 3, making a generic translation on both x and y. Let us write $x - y = d(x_1 - y_1)$ with suitable points x_1, y_1. This can be done by Corollary 1 of Theorem 3. The points x_1, y_1 will also be generic points of J, and $Z(x_1), Z(y_1)$ are defined. According to Theorem 4 and Corollary 2 of Theorem 4, Chapter III, § 3 we have

$$Z(x) - Z(y) \sim d[Z(x_1) - Z(y_1)] \sim W_u - W$$

with a suitable point u. This shows that φ_W is surjective, and concludes the proof.

COROLLARY 2. *Let X be a non-degenerate divisor. Then the homomorphism*

$$\varphi_X : A \to \mathrm{Pic}\,(A)$$

is surjective. If A is defined over k, we can find a positive non-degenerate divisor which is rational over k.

Proof: The first assertion is obvious from the preceding corollary. As to the second, it is merely a repetition of a remark to the effect that we can find a projective embedding of A defined over k, so that a hyperplane section rational over k is non-degenerate.

We recall that a divisor is said to be a torsion divisor if some multiple of it is algebraically equivalent to 0.

COROLLARY 3. *A divisor X on an abelian variety is a torsion divisor if and only if $X \equiv 0$.*

Proof: If $X \equiv 0$, we have just seen that dX is algebraically equivalent to 0. Conversely, assume that dX is in $D_a(A)$. According to Proposition 4 of Chapter III, § 3 we have $\varphi_{dX} = 0$. But $\varphi_{dX} = d \cdot \varphi_X$, and for u generic on A, du is also generic on A according to Corollary 1 of Theorem 3. Hence

$$\varphi_{dX}(u) = \varphi_X(du) = 0,$$

and $\varphi_X = 0$, $X \equiv 0$.

As mentioned in the last chapter, it will be shown in Chapter

VII that if X is a torsion divisor, then there exists a power p^m of the characteristic such that $p^m X$ is algebraically equivalent to 0. It is in fact true that the torsion group is trivial (Barsotti [7]) but we shall neither prove nor use this fact in this book.

As another application of Theorem 4 let us show that the torsion divisors on an arbitrary variety can be put as direct summands of the Picard group.

COROLLARY 4. *Let V be any variety, and let X be a divisor on V such that $mX \in D_a(V)$ for some integer $m \neq 0$. Then there exists a divisor Y algebraically equivalent to X such that mY is linearly equivalent to 0.*

Proof: There exists an abelian variety J and a divisor Z on $J \times V$ such that $mX = Z(x) - Z(y)$ with two suitable points x, y of J. As in Corollary 1, we may assume them generic after making a generic translation on J, because we have

$$Z(x+u) - Z(y+u) \sim Z(x) - Z(y).$$

Let us write $x - y = m(x_1 - y_1)$ with two generic points x_1, y_1. Then

$$m[Z(x_1) - Z(y_1)] \sim Z(x) - Z(y) \sim mX.$$

The divisor $Y = X - Z(x_1) + Z(y_1)$ satisfies the requirements of our corollary.

The following sections, on numerical equivalence and on the Picard variety are independent of each other, and we could now treat immediately the Picard variety. The order in which they occur has been selected randomly.

§ 3. *Numerical equivalence*

Let V^r be a variety, complete and non-singular. Let Z be a cycle on V of dimension s, and let us denote the group of such cycles by $Z(V, s)$. We say that Z is *numerically equivalent to* 0, and write $Z \approx 0$, if for every cycle Y of complementary dimension (i.e., of dimension $r - s$) such that $Z \cdot Y$ is defined we have

$$\deg (Z \cdot Y) = 0.$$

We say that two cycles Z, Z' are numerically equivalent if $Z - Z' \approxeq 0$. It can be shown that the cycles that are numerically equivalent to 0 form a subgroup denoted by $Z_n(V, s)$. The factor group is denoted by $N_s(V)$. On an abelian variety, one can develop certain lemmas of a formal nature concerning numerical equivalence more easily than on an arbitrary variety. We shall therefore restrict ourselves to this case: We assume $V = A$ is an abelian variety.

Let us begin by showing that the cycles $Z \approxeq 0$ form a group. Suppose that $Z \cdot Y$ is defined. According to Proposition 5 of Chapter I, § 2 if u is a generic point of A then $Z \cdot Y_u$ is also defined. Proposition 4 of Chapter I, § 2 together with the principle of conservation of number (F-VII$_6$ Th. 13) show that

$$\deg (Z \cdot Y) = \deg (Z \cdot Y_u).$$

Similarly, for u generic, we have $\deg (Z_u \cdot Y) = \deg (Z \cdot Y)$. This shows that Z is numerically equivalent to a generic translation of itself. Let Z, Z' be two cycles of dimension s on A, such that $(Z + Z') \cdot Y$ is defined. If u is generic on A then $(Z_u + Z'_u) \cdot Y$, $Z_u \cdot Y$, and $Z'_u \cdot Y$ are defined. The preceding remark shows that $\deg [(Z + Z') \cdot Y] = 0$ if both Z and Z' are numerically equivalent to 0. Hence we see that the cycles of dimension s which are numerically equivalent to 0 form a subgroup of $Z(V, s)$.

Let us form the graded module

$$N_*(A) = \sum_{i=0}^{r} N_i(A)$$

which is the direct sum of the $N_i(A)$. We are going to define a ring structure on this module by the intersection product. Let X_1, X_2 be two cycles of dimension i, and Y_1, Y_2 two cycles of dimension j on A. After making generic translations, we see by an argument similar to the preceding one that if $X_1 \approxeq X_2$ and $Y_1 \approxeq Y_2$, and if $X_1 \cdot Y_1$ and $X_2 \cdot Y_2$ are defined, then $X_1 \cdot Y_1 \approxeq X_2 \cdot Y_2$. Furthermore, if $\xi \in N_i(A)$ and $\eta \in N_j(A)$, we can find two cycles X, Y representing ξ, η respectively such that $X \cdot Y$ is defined: it suffices again to make a generic translation on arbitrarily selected representatives. We can therefore define the product $\xi \cdot \eta$ to be that element of $N_{i+j-r}(A)$ represented by $X \cdot Y$. It is then clear that this product defines a ring structure on $N_*(A)$.

We note that in dimensions 0 and r, the modules $N_0(A)$ and $N_r(A)$ are infinite cyclic groups, generated by a canonical element. In the case of $N_0(A)$, a canonical generator ξ is the class of a point. In the case of $N_r(A)$, it is the class of A. In both cases, if ξ is this canonical generator, and $\eta = m \cdot \xi$, we shall say that m is the *degree* of η and we write

$$m = \deg(\eta).$$

Note that if $i + j = r$, we have a bilinear map

$$N_i(A) \times N_{r-i}(A) \to \mathbf{Z}$$

defined by $(\xi, \eta) \to \deg(\xi \cdot \eta)$.

Let A, B be two abelian varieties of the same dimension, and let $\alpha : A \to B$ be a homomorphism. Then α induces a contravariant ring homomorphism

$$\alpha^{-1} : N_*(B) \to N_*(A).$$

In order to see this, let us take an element $\eta \in N_i(B)$, and let Y be a representative cycle of η. After a generic translation, we can always choose Y such that $\alpha^{-1}(Y)$ is defined. The numerical equivalence class of $\alpha^{-1}(Y)$ is then an element of $N_i(A)$ because we have assumed $\dim A = \dim B$.

Let Y_1, Y_2 be two representatives of η such that $\alpha^{-1}(Y_1)$ and $\alpha^{-1}(Y_2)$ are defined. We must prove that $\alpha^{-1}(Y_1) \approx \alpha^{-1}(Y_2)$, or in other words, putting $Y = Y_1 - Y_2$, we must prove that if $Y \approx 0$ on B and $\alpha^{-1}(Y)$ is defined, then $\alpha^{-1}(Y) \approx 0$ on A. Let X be a cycle of complementary dimension $r - i$ on A, such that $\alpha^{-1}(Y) \cdot X$ is defined. Let u be a generic point of A. We know that

$$\deg(\alpha^{-1}(Y) \cdot X) = \deg(\alpha^{-1}(Y) \cdot X_u).$$

It will suffice to prove that the right-hand side of this equation is equal to 0.

The degree of a 0-cycle being preserved under the operation pr, it will therefore suffice to prove that the degree of the cycle

$$[\Gamma_\alpha \cdot (A \times Y)] \cdot (A \times X_u) = [(A \times Y) \cdot \Gamma_\alpha] \cdot (A \times X_u) \quad (1)$$

is equal to 0 (F-VII_6 Th. 16 and the definitions). Since we have made a generic translation, all the desirable intersections are defined, we can use associativity, and the above cycle is equal to

$$(A \times Y) \cdot [\Gamma_\alpha \cdot (A \times X_u)].$$

If we now take a projection pr relative to the first factor, and if we use again F-VII_6 Th. 16 then we find the cycle

$$Y \cdot \nu(\alpha) X_u.$$

The degree of this cycle is equal to 0 by the hypothesis that $Y \approx 0$.

We have just shown that $\alpha^{-1} : N_i(B) \to N_i(A)$ is a homomorphism for the additive structure. It is in fact a ring homomorphism on $N_*(A)$, in view of the formula

$$\alpha^{-1}(Y_1 \cdot Y_2) = \alpha^{-1}(Y_1) \cdot \alpha^{-1}(Y_2)$$

whenever these intersections are defined (Appendix § 1, Theorem 4). As before, we can always find representatives Y_1, Y_2 of two given classes η_1, η_2 such that these intersections are defined.

Let X be a cycle on A. Then $\Gamma_\alpha \cdot (X \times A)$ is defined, and we shall consider the image $\alpha(X)$ in the sense of intersection theory, namely we put as usual,

$$\alpha(X) = \mathrm{pr}_2 [\Gamma_\alpha \cdot (X \times A)].$$

It is a cycle on B of the same dimension as X. An argument similar to the preceding one shows that if ξ is an element of $N_i(A)$ and X a representative of ξ, then the class of $\alpha(X)$ in $N_i(B)$ does not depend on the selected representative X. We denote it by $\alpha(\xi)$. One sees immediately that the map $\xi \to \alpha(\xi)$ gives an additive homomorphism

$$\alpha : N_i(A) \to N_i(B).$$

It is not however a ring homomorphism for the intersection product.

If X is a cycle on A, and Y a cycle on B such that $\dim X = \dim Y = r$, then from (1) we see immediately that

$$\deg (\alpha(X) \cdot Y) = \deg (X \cdot \alpha^{-1}(Y)) \tag{2}$$

each time that the intersections occurring in this formula are defined. We shall call (2) the *transposition formula.* It shows that if α is an endomorphism of A, then the additive homomorphisms induced by α on $N_*(A)$ are the transpose of each other.

Finally, if we want to define a direct homomorphism

$$\alpha : N_*(A) \to N_*(B)$$

as a ring homomorphism, we must take the Pontrjagin product of Chapter I, § 2. We know that the Pontrjagin product $X_1 * X_2$ is defined between any two cycles of A. We are going to show that it defines a ring structure on $N_*(A)$. It amounts to showing that if $X \approx 0$ and Y is any cycle, then $X * Y \approx 0$. According to the definitions, and Proposition 3 of Chapter I, § 2 we must show that the degree of the cycle

$$\mathrm{pr}_3 [S_2 \cdot (X \times Y \times A)] \cdot Z$$

is equal to 0 for any cycle Z such that this intersection is defined. We may assume that $\dim X + \dim Y \leq \dim A$, and we can replace Z by a generic translation Z_u. Since the degree of a 0-cycle is preserved under the operation pr, we see from F-VII$_6$ Th. 16 that it suffices to prove that the degree of the cycle

$$[S_2 \cdot (X \times Y \times A)] \cdot (A \times A \times Z_u) = [(X \times Y \times A) \cdot S_2] \cdot (A \times A \times Z_u)$$

is equal to 0. We can apply the associativity theorem, and the identity $(X \times Y \times A) = (X \times A \times A) \cdot (A \times Y \times A)$ to transform this cycle into

$$[(X \times A \times A) \cdot (A \times Y \times A)] \cdot [S_2 \cdot (A \times A \times Z_u)].$$

On the other hand, $(A \times Y \times A) \cdot S_2$ is defined, and we can apply Proposition 17 of F-VII$_6$ together with the associativity theorem to the three cycles $(A \times Y \times A)$, S_2, $(A \times A \times Z_u)$. Our cycle is therefore equal to

$$(X \times A \times A) \cdot W$$

where $W = (A \times Y \times A) \cdot S_2 \cdot (A \times A \times Z_u)$. If we take the projection on the first factor, we get

$$\mathrm{pr}_1 [(X \times A \times A) \cdot W] = X \cdot \mathrm{pr}_1 W$$

and the degree of this cycle is equal to 0 by the hypothesis $X \approx 0$. We have thus shown that the Pontrjagin product induces a ring structure on $N_*(A)$.

Let us return to the homomorphism $\alpha : A \to B$. Then $\alpha(X)$ is defined for every cycle X of A, and it gives rise to a ring homomorphism on the cycles, under the Pontrjagin product, in view of Chapter I, § 2. We must therefore show that it preserves numerical equivalence. This is a consequence of an argument entirely similar to the one used above to show that α^{-1} preserves numerical equivalence. Indeed, suppose $X \approx 0$ on A, and let Y be a cycle of complementary dimension on B such that $\alpha(X) \cdot Y$ is defined. Let v be a generic point of B. Then

$$\deg (\alpha(X) \cdot Y) = \deg (\alpha(X) \cdot Y_v),$$

and we must show that this degree is equal to 0. By definition, and by the projection theorem (F-VII$_6$ Th. 16) we have

$$\begin{aligned} \mathrm{pr}_2 [\Gamma_\alpha \cdot (X \times B)] \cdot Y_v &= \mathrm{pr}_2 \{[\Gamma_\alpha \cdot (X \times B)] \cdot (A \times Y_v)\} \\ &= \mathrm{pr}_2 \{[(X \times B) \cdot \Gamma_\alpha] \cdot (A \times Y_v)\}. \end{aligned}$$

Associativity, and the fact that the degree of a 0-cycle is preserved under projection, brings us back to showing that the degree of $(X \times B) \cdot [\Gamma_\alpha \cdot (A \times Y_v)]$ is equal to 0. One sees this immediately by using the hypothesis $X \approx 0$, and by projecting on the first factor.

Summarizing, we can express the results obtained above in the following manner.

THEOREM 5. *Let A, B be two abelian varieties of the same dimension r, and let $N_*(A)$, $N_*(B)$ be the graded modules obtained from numerical equivalence. Let $\alpha : A \to B$ be a homomorphism. Then α induces a contravariant ring homomorphism for the intersection product*

$$\alpha^{-1} : N_*(B) \to N_*(A),$$

and induces a covariant ring homomorphism for the Pontrjagin product,

$$\alpha : N_*(A) \to N_*(B).$$

If $\xi \in N_i(A)$ *and if* $\eta \in N_{r-i}(B)$, *then we have*

$$\deg(\alpha(\xi) \cdot \eta) = \deg(\xi \cdot \alpha^{-1}(\eta)).$$

We can reformulate in our new language the result of Proposition 6, Chapter I, § 2. If ξ is an element of $N_i(A)$, and X a cycle representing ξ, then X^- represents a class in $N_i(A)$ which will be denoted by ξ^-.

PROPOSITION 3. *If* $\xi \in N_i(A)$ *and* $\eta \in N_{r-i}(A)$, *then*

$$\deg(\xi \cdot \eta^-) = \deg(\xi^- \cdot \eta) = \deg(\xi * \eta).$$

In order to obtain numerical results on A, we also need the following general fact.

PROPOSITION 4. *Let X be a torsion divisor on the abelian variety A (in particular, X may be algebraically equivalent to* 0). *Then X is numerically equivalent to* 0. *Hence if* $X \equiv 0$ *then* $X \approx 0$.

Proof: The last assertion comes from Theorem 4 of § 2. Suppose now that X is algebraically equivalent to 0, and that we have written $X = Z(P) - Z(Q)$, with two simple points P, Q on a parameter variety, and a cycle Z on the product. Let Y be a cycle of complementary dimension such that $X \cdot Y$ is defined, and let u be a generic point of A. Then we know that $\deg(X \cdot Y) = \deg(X \cdot Y_u)$. Furthermore, $Z(P) \cdot Y_u$ and $Z(Q) \cdot Y_u$ are defined, and it suffices to show that they have the same degree. We may assume that Z is a positive cycle, by linearity, and using the fact that for u generic, all the intersections with which we are concerned will be defined. The result that we are seeking is then a consequence of the compatibility of the intersection with the specialization of cycles. It would also be easy (but tedious) to reduce our proof to F-VII$_6$ Th. 13. We have thus shown that algebraic equivalence implies numerical equivalence. If X is a torsion divisor, it is then clear that X is numerically equivalent to 0. This concludes the proof.

Having seen that if $X \equiv 0$ then X is numerically equivalent to 0, we obtain a canonical homomorphism

$$N'(A) \rightarrow N_{r-1}(A).$$

We shall prove in the next chapter that in fact this homomorphism is an isomorphism: A divisor X is numerically equivalent to 0 if and only if $X \equiv 0$. For the moment, we need only the result just stated. We observe that an equality between elements of $N'(A)$ gives rise to an equality between elements of $N_{r-1}(A)$. We shall apply this remark to the formulas of Proposition 2, § 1.

A divisor X on an abelian variety is said to be a *polar divisor* if there exists a positive integer m such that mX is ample, i.e., linearly equivalent to a hyperplane section in some projective embedding. For the rest of this section, we shall assume that ξ is an element of $N_{r-1}(A)$ having a representative which is a polar divisor, and we shall denote by X a representative of ξ. We note that if Z is a positive cycle of dimension 1 such that $X \cdot Z$ is defined, then $\deg (X \cdot Z) > 0$. We then see immediately that if we raise ξ to the rth power in the numerical intersection ring, then we get $\deg (\xi^r) > 0$.

From Proposition 2 of § 1, we get

$$(n\delta)^{-1}(\xi) = n^2 \cdot \xi,$$

this equality being now viewed as an equality between elements of $N_{r-1}(A)$ (and not any more of $N'(A)$). If we raise both sides of this equation to the rth power (for the intersection product) we find

$$\nu(n\delta) \cdot \deg (\xi^r) = n^{2r} \cdot \deg (\xi^r)$$

taking into account the fact that if η is an element of $N_0(A)$ then

$$\deg (\alpha^{-1}(\eta)) = \nu(\alpha) \cdot \deg (\eta).$$

If n is not divisible by the characteristic p, this shows that the degree of $n\delta$ is prime to p, and that $n\delta$ is separable. The number of points in the kernel of $n\delta$ is therefore equal to n^{2r}.

More generally, if $\alpha_1, \ldots, \alpha_d$ are endomorphisms of A and $m_1, \ldots, m_d$ are integers, then we have

$$(m_1\alpha_1 + \ldots + m_d\alpha_d)^{-1}(\xi) = \tfrac{1}{2} \sum m_i m_j D_\xi(\alpha_i, \alpha_j).$$

Raising both sides of this equation to the rth power, we find that $\nu(m_1\alpha_1 + \ldots + m_d\alpha_d)$ as a function of $(m_1, \ldots, m_d)$ is a homogeneous polynomial of degree $2r$ with rational coefficients which one can easily determine explicitly in terms of intersections of the $D_\xi(\alpha_i, \alpha_j)$.

If the α_i are homomorphisms of one abelian variety A into another B of the same dimension, then either A and B are not isogenous, in which case $\nu(m_1\alpha_1 + \ldots + m_d\alpha_d) = 0$ for all m_i, or there exists an isogeny $\lambda : B \to A$, that is to say a homomorphism of finite degree. If we put $\beta_i = \lambda\alpha_i$, the β_i are endomorphisms of A to which we can apply the preceding remark. Since $\nu(m_1\alpha_1 + \ldots + m_d\alpha_d) = (1/\nu(\lambda))\,\nu(m_1\beta_1 + \ldots + m_d\beta_d)$ we see that $\nu(m_1\alpha_1 + \ldots + m_d\alpha_d)$ is also a homogeneous polynomial of degree $2r$ with rational coefficients.

We have therefore proved the following theorem.

THEOREM 6. *Let $\alpha_1, \ldots, \alpha_d : A^r \to B^r$ be homomorphisms of an abelian variety into another of the same dimension, and let $m_1, \ldots, m_d$ be integers. Then $\nu(m_1\alpha_1 + \ldots + m_d\alpha_d)$ is a homogeneous polynomial of degree $2r$ with rational coefficients as a function of the m_i. In particular, $\nu(n\delta) = n^{2r}$. If n is prime to the characteristic, then the kernel of $n\delta$ has exactly n^{2r} elements.*

Let α be an endomorphism of A. We have the symbol

$$D_\xi(\alpha) = D_\xi(\alpha, \delta) = (\alpha + \delta)^{-1}(\xi) - \alpha^{-1}(\xi) - \xi$$

for each element ξ of $N_{r-1}(A)$. It is therefore also an element of $N_{r-1}(A)$. We can write

$$(\alpha + n\delta)^{-1}(\xi) = n^2 \cdot \xi + n \cdot D_\xi(\alpha) + \alpha^{-1}(\xi).$$

Raising this equation to the rth power, we find

$$\nu(\alpha + n\delta) \cdot \deg(\xi^r) = c_{2r}n^{2r} + c_{2r-1}n^{2r-1} + \ldots + c_0$$

where the c_i are rational integers which are easily determined explicitly in terms of intersections of $\alpha^{-1}(\xi)$, ξ, and $D_\xi(\alpha)$. The most important coefficients are

$$c_{2r} = \deg(\xi^r), \quad c_{2r-1} = r \cdot \deg(\xi^{r-1} \cdot D_\xi(\alpha)), \quad c_0 = \nu(\alpha) \cdot \deg(\xi^r).$$

We see therefore that $\nu(\alpha + n\delta)$ is a polynomial in n with rational coefficients. In order to be consistent with the theory of matrix representation, we shall call the polynomial $P(t)$ such that $P(n) = \nu(\alpha - n\delta)$ the *characteristic polynomial* of α. The coefficient of the term of degree $2r - 1$ in $\nu(\alpha + n\delta)$ will be called the *trace* of α and will be written $\operatorname{tr}(\alpha)$. We have the following fundamental formula:

$$\operatorname{tr}(\alpha) = \frac{r}{\deg(\xi^r)} \deg(\xi^{r-1} \cdot D_\xi(\alpha)),$$

where we recall that ξ is an element of $N_{r-1}(A)$ containing a polar divisor X, thus guaranteeing that $\deg(\xi^r) > 0$. We shall return later to a study of the trace, when we have defined suitable involutions on the algebra of endomorphisms of A. For this we shall need the formalism of the Picard variety. We shall also show in Chapter VII that the characteristic polynomial as we have defined it here coincides with the characteristic polynomial obtained from an l-adic representation.

We can dualize the preceding formula by using cycles of dimension 1 instead of divisors, and the Pontrjagin product instead of the intersection product in the numerical equivalence classes. We proceed as follows.

Let ζ be an element of $N_1(A)$, and let α, β be two homomorphisms of A into an abelian variety B of the same dimension. We can define the class

$$Z_\zeta(\alpha, \beta) = (\alpha + \beta)(\zeta) - \alpha(\zeta) - \beta(\zeta)$$

and

$$Z_\zeta(\alpha) = Z_\zeta(\alpha, \delta).$$

They are elements of $N_1(B)$. The transposition formula of Theorem 5 gives us

$$\deg(Z_\zeta(\alpha, \beta) \cdot \eta) = \deg(\zeta \cdot D_\eta(\alpha, \beta))$$

for $\eta \in N_{r-1}(B)$. This allows us to transport to $N_1(A)$ the formulas in Proposition 2 of § 1. In particular, we obtain

$$(\alpha + n\delta)(\zeta) = n^2 \cdot \zeta + n \cdot Z_\zeta(\alpha) + \alpha(\zeta).$$

Let us take $A = B$, and take for ζ an element of $N_1(A)$ which contains a curve (irreducible) of A which generates A. Then $\zeta^{*r} \neq 0$ and $\deg(\zeta^{*r}) > 0$ if we denote by ζ^{*s} the Pontrjagin product of ζ with itself taken s times. Consequently, if we raise each side of the above equation to the rth power, the product being this time the Pontrjagin product, we obtain for the trace of an endomorphism α the formula

$$\operatorname{tr}(\alpha) = \frac{r}{\deg(\zeta^{*r})} \deg(\zeta^{*(r-1)} * Z_\xi(\alpha)).$$

Taking into account Proposition 3, the corollary to Proposition 2 of § 1, and Proposition 4 we can rewrite this formula in the following manner:

$$\operatorname{tr}(\alpha) = \frac{r}{\deg(\zeta^{*r})} \deg(\zeta^{*(r-1)} \cdot Z_\xi(\alpha)).$$

Summarizing, we have therefore proved the following theorem.

THEOREM 7. *Let α be an endomorphism of A. Then $\nu(\alpha + n\delta)$ as a function of n is a polynomial with rational coefficients, of type*

$$\nu(\alpha + n\delta) = n^{2r} + \operatorname{tr}(\alpha) n^{2r-1} + \ldots + \nu(\alpha).$$

If ξ is an element of $N_{r-1}(A)$ which contains a polar divisor, and ζ is an element of $N_1(A)$ which contains a curve generating A, then $\operatorname{tr}(\alpha)$ is given by the formulas

$$\operatorname{tr}(\alpha) = \frac{r}{\deg(\xi^r)} \deg(\xi^{r-1} \cdot D_\xi(\alpha)) = \frac{r}{\deg(\zeta^{*r})} \deg(\zeta^{*(r-1)} \cdot Z_\xi(\alpha)).$$

In particular, let us apply this to the case where $A = J$ is the Jacobian of a curve C, and ζ is the numerical equivalence class of the curve itself, which we may assumed contained in its Jacobian by Proposition 4 of Chapter II, § 2. The corollary of this proposition gives us explicitly ζ^{*r} and $\zeta^{*(r-1)}$. In the case of Jacobians, the formulas for the trace can therefore be simplified as follows.

COROLLARY. *Let J be the Jacobian of a complete non-singular curve C which we may assumed contained in J under the canonical mapping. Let θ be the class of Θ in $N_{r-1}(J)$, and ζ the class of C in*

$N_1(J)$. *Then for an endomorphism* α *of* J *we have*

$$\operatorname{tr}(\alpha) = \deg (D_\theta(\alpha) \cdot \zeta).$$

As in Chapter II, let us denote by $\operatorname{End}_{\mathbf{Q}}(A)$ the tensor product of $\operatorname{End}(A)$ with the rational numbers. We may also take the tensor product $N_*(A)$ with the rationals, denoting it by $N_{*\mathbf{Q}}(A)$. We can extend to it the intersection and Pontrjagin products.

Furthermore, we can extend the definition of the trace to elements of $\operatorname{End}_{\mathbf{Q}}(A)$, and also the definition of the symbol $\nu(\alpha)$, by putting $\nu(\alpha/n) = n^{-2r}\nu(\alpha)$ for $\alpha \in H(A, B)$ and n an integer $\neq 0$. All the results which we have obtained, and in particular those of Theorems 6 and 7 can then be extended to arbitrary elements of $H_{\mathbf{Q}}(A, B)$ and $\operatorname{End}_{\mathbf{Q}}(A)$. An element of $N_{r-1}(A) \otimes \mathbf{Q}$ is said to be a *polar class* if a suitable multiple of it contains a polar divisor. In Theorem 7, we can use for ξ a polar class. An analogous remark holds for ζ.

The characteristic polynomial of an element $\alpha \in \operatorname{End}_{\mathbf{Q}}(A)$ is also defined by $\nu(\alpha - n\delta)$, and its roots will be called the *characteristic roots* of α. If ω_i $(i = 1, \ldots, 2r)$ are these roots, we have

$$\nu(\alpha) = \prod_{i=1}^{2r} \omega_i\,, \qquad \operatorname{tr}(\alpha) = \sum_{i=1}^{2r} \omega_i\,.$$

We are now going to show that if $F(T) = a_d T^d + \ldots + a_0$ is a polynomial with rational coefficients, and if we denote by $F(\alpha)$ the element $a_d \alpha^d + \ldots + a_0$ of $\operatorname{End}_{\mathbf{Q}}(A)$, then the characteristic roots of $F(\alpha)$ behave exactly like the characteristic roots of a linear transformation. In fact this result will be used later to show that our characteristic roots coincide with those of an l-adic representation of $\operatorname{End}_{\mathbf{Q}}(A)$.

THEOREM 8. *Let* α *be an element of* $\operatorname{End}_{\mathbf{Q}}(A)$, *and* $\omega_1, \ldots, \omega_{2r}$ *its characteristic roots. Let* $F(T)$ *be a polynomial with rational coefficients. Then the characteristic roots of* $F(\alpha)$ *are* $F(\omega_1), \ldots, F(\omega_{2r})$. *Furthermore, we have*

$$\nu(F(\alpha)) = \prod F(\omega_i), \quad \text{and} \quad \operatorname{tr}(F(\alpha)) = \sum F(\omega_i).$$

Proof: We may assume without loss of generality that the leading coefficient of $F(T)$ is equal to 1, that is to say,

$$F(T) = T^d + a_{d-1}T^{d-1} + \dots + a_0$$

with rational numbers a_i. According to Theorem 6, there exists a polynomial $P(a_0, \dots, a_{d-1})$ of degree $2r$ with rational coefficients such that $\nu(F(\alpha)) = P(a)$. Put

$$Q(a) = \prod_{i=1}^{2r} F(\omega_i).$$

Then $Q(a)$ is also a polynomial of degree $2r$ in $a_0, \dots, a_{d-1}$ with rational coefficients. We must show that $P(a) = Q(a)$. Suppose that we have a polynomial $G(T)$ of type

$$G(T) = \prod_{j=1}^{d} (T + x_j) = T^d + b_{d-1}T^{d-1} + \dots + b_0$$

with rational numbers x_j. Then

$$\nu(G(\alpha)) = \prod_{j=1}^{d} \nu(\alpha + x_j)$$

and consequently, according to the definition of the characteristic polynomial and the characteristic roots, we get

$$\nu(G(\alpha)) = \prod_{i=1}^{2r}\prod_{j=1}^{d} (\omega_i + x_j) = \prod_{i=1}^{2r} G(\omega_i) = Q(b).$$

Since on the other hand we have $\nu(G(\alpha)) = P(b)$ by definition of P, it follows that $P(b) = Q(b)$ whenever the numbers (b) are defined as above, no matter what rational numbers x_j we select. It follows that if we substitute for the (b) in the polynomial $P(b) - Q(b)$ their expressions in terms of $x_1, \dots, x_d$ we obtain a polynomial in these d variables which vanishes for all rational values of these variables, and hence vanishes identically. If we denote by $y_1, \dots, y_d$ the roots of $F(T) = T^d + a_{d-1}T^{d-1} + \dots + a_0$ and if we substitute for the (a) in the polynomial $P(a) - Q(a)$ their expressions as functions of the y_i we obtain 0. We have therefore $P(a) = Q(a)$ for all (a), that is to say

$$\nu(F(\alpha)) = \prod F(\omega_i).$$

The assertion concerning the trace follows now immediately by considering $\nu(F(\alpha) + n\delta)$ with an arbitrary integer n.

REMARK. The proof which we have just given can be axiomatized in the following manner. We suppose given an algebra R over the rationals (possibly infinite dimensional) and a function $\nu(\alpha)$ with values in $\mathbf{Q}$ satisfying the following conditions:

(i) $\nu(\alpha\beta) = \nu(\alpha)\nu(\beta)$;

(ii) for all sets of rational numbers $m_1, \ldots, m_d$ and for given elements $\alpha_1, \ldots, \alpha_d$ of R, the function

$$\nu(m_1\alpha_1 + \ldots + m_d\alpha_d)$$

is a homogeneous polynomial of degree s in the m_i, such that in particular we have $\nu(m\alpha) = m^s\nu(\alpha)$.

Then the roots of the polynomial $\nu(\alpha + n\delta)$, as functions of n, satisfy the condition stated in Theorem 8.

§ 4. *The Picard variety of an abelian variety*

Let V be a complete variety, non-singular in codimension 1, defined over a field k. A *Picard variety of* V, *defined over* k, is a couple (B, D) consisting of an abelian variety B defined over k, and a divisor D on the product $V \times B$, rational over k, and called a *Poincaré divisor*, satisfying the following conditions:

(1) *For every point* $w \in B$ *and every generic point* $z \in B$ *over* $k(w)$, *the mapping*

$$w \to \mathrm{Cl}\,[{}^tD(z + w) - {}^tD(z)]$$

is an isomorphism of B *onto the Picard group* $\mathrm{Pic}\,(V) = D_a(V)/D_l(V)$.

(The theorem of the square guarantees that the linear equivalence class above does not depend on the choice of auxiliary generic point z selected.)

(2) *If* ${}^tD(z + w) - {}^tD(z)$ *is linearly equivalent to a divisor* X *on* V, *and if* X *is rational over a field* $K \supset k$, *then the point* w *in condition* (1) *is rational over* K.

From now on, we shall use freely the theorem of the square without making further references to it.

It is obvious that if k' is a field containing k, and if (B, D) is a

Picard variety of V defined over k, then it is also a Picard variety of V defined over k'.

From the two conditions (1) and (2) we can deduce the uniqueness of the Picard variety. Let us recall that a divisor on a product $V \times W$ is said to be trivial if it is of type $X \times W + V \times Y + (\varphi)$ for some divisor X on V, Y on W, and a function φ on $V \times W$.

(3) *If (B', D') is another Picard variety of V defined over k, then B' is birationally isomorphic to B over k. The Poincaré divisor is uniquely determined up to a trivial divisor, namely, if we put $B = B'$, then we may write*

$$D' = D + (\varphi) + X \times B + V \times Y$$

for some divisors X on V and Y on B, both rational over k, and a function φ on the product, defined over k.

Indeed, for a given linear equivalence class of the Picard group, associated with a point x on B and x' on B' we may write

$${}^tD(x + z) - {}^tD(z) \sim {}^tD'(x' + z') - {}^tD'(z')$$

with a generic point $z' \in B'$ over $k(x')$ and a generic point $z \in B$ over $k(x)$. Condition (2) shows that x' is rational over $k(x, z)$. Since we can replace z by another generic point z_1 of B, independent of z over $k(x)$, we see that x' is rational over $k(x, z) \cap k(x, z_1) = k(x)$. By symmetry, we must have $k(x) = k(x')$, and the correspondence $x \to x'$ obviously gives a one-one birational correspondence between B and B', and hence a birational isomorphism between them.

To prove the uniqueness of D, we may suppose $B = B'$. Let us observe right away that if D, D' differ by a trivial divisor rational over k, then they give the same isomorphism of B on Pic (V) as in (1). Conversely, put $E = D - D'$. Then for any generic point z of B over k, the linear equivalence class ${}^tE(z)$ of V is constant, by the very definition of E. The seesaw principle shows that E is trivial (Appendix, § 2, Theorem 6) and the corollary to that theorem gives us the rationality statements, in view of the fact that B has a simple rational point, the origin, and that V is assumed complete, non-singular in codimension 1.

The uniqueness of the Picard variety allows us to identify the class Cl (X) of a divisor algebraically equivalent to 0 on V with the point on the Picard variety associated with it under the isomorphism of condition (1). We may therefore also write $B =$ Pic (V). In the cases where we shall still have to distinguish between the abelian variety B and the linear equivalence class of X, the context will always make our meaning clear.

The next proposition gives us a converse for (2).

PROPOSITION 5. *Let V be complete, non-singular in codimension* 1, *defined over k. Let (B, D) be a Picard variety of V, also defined over k, and let w be a point of B. Then there exists a Poincaré divisor D' on $V \times B$ rational over k such that ${}^tD'(w)$ is defined, and there exists a divisor $X \in D_a(V)$ rational over $k(w)$ such that* Cl $(X) = w$.

Proof: This is essentially a special case of the corollaries to Propositions 4 and 5 of the Appendix, § 2.

In particular, we see that the field $k(w)$ is equal to the intersection of all fields of rationality for divisors $X \in D_a(V)$ such that Cl $(X) = w$.

It will now be useful to give other conditions characterizing the Picard variety, for instance the following ones in which only generic points of B occur.

THEOREM 9. *Let V be a complete variety, non-singular in codimension* 1 *and defined over a field k. Let B be an abelian variety defined over k and D a divisor on $V \times B$ rational over k, having the following properties:*

(i) *If z is a generic point of B over k, then $k(z)$ is the smallest field of rationality of ${}^tD(z)$.*

(ii) *If z is a generic point of B over k, and if there exists a divisor X of V rational over a field $K \supset k$ such that ${}^tD(z) \sim X$, then z is rational over K.*

(iii) *If z, w are two generic points of B over k (not necessarily independent) then ${}^tD(z) \sim {}^tD(w)$ if and only if $z = w$.*

(iv) *For every $X \in D_a(V)$ and every field $K \supset k$ there exist two generic points z, w of B over K such that*

$$X \sim {}^tD(z) - {}^tD(w).$$

Then (B, D) is a Picard variety of V, defined over k, and conversely, every Picard variety of V defined over k has properties (i)—(iv) *above.*

Proof: Suppose first that (B, D) is a Picard variety defined over k. Then our present condition (iv) is essentially the same as (1) in the definition of the Picard variety, taking into account the theorem of the square, and the fact that we can always add a generic point to the parameters without changing the linear equivalence class. To prove (i), let K be the smallest field of rationality of ${}^tD(z)$ containing k. If w is a generic point of B independent of z, then

$$ {}^tD(z+w) - {}^tD(w) \sim {}^tD(z) - {}^tD(0) $$

if ${}^tD(0)$ is defined. This can always be achieved, by Proposition 5. The divisor ${}^tD(z) - {}^tD(0)$ is rational over K, and hence by the definition of the Picard variety, z is rational over K. The same argument can be applied to prove (ii). Condition (iii) comes from the fact that the parametrization of the divisor classes by the Picard variety B is one-one.

Conversely, suppose that (B, D) satisfies (i)—(iv). Then (iv) shows that the mapping

$$ w \to \mathrm{Cl}\,[{}^tD(z+w) - {}^tD(z)] $$

is a homomorphism of B onto the Picard group Pic (V), and (iii) shows that in fact it is an isomorphism. Suppose that X is rational over $K \supset k$, and that $X \sim {}^tD(z+x) - {}^tD(z)$ with z generic over $K(x)$. We get ${}^tD(z+x) \sim {}^tD(z) + X$, and hence by (ii), $z+x$ is rational over $K(z)$, so that x is rational over $K(z)$. If z_1, z_2 are two independent generic points of B over $K(x)$, it follows that x is rational over $K(z_1) \cap K(z_2) = K$, which is condition (2) in the definition of the Picard variety. This concludes the proof.

THEOREM 10. *Let A be an abelian variety, defined over k. Then there exists an abelian variety $\hat{A}$ defined over k and a positive divisor D on $A \times \hat{A}$ rational over k such that $(\hat{A}, D)$ is a Picard variety of A defined over k. Furthermore, A and $\hat{A}$ are isogenous.*

Proof: We merely reap without further pains the fruits of our labors, using on the one hand Corollary 2 of Theorem 4, § 2 together with Theorem 5 of Chapter III, § 4. If X is a positive non-degenerate divisor on A, rational over k, we take $s_2^{-1}(X)$ on $A \times A$ and pass to the quotient by means of the second theorem just quoted.

It is convenient to insert at this point the fact that the Picard variety of a product is the product of the Picard varieties. We have first a set-theoretic result.

PROPOSITION 6. *Let V, W be two arbitrary varieties, and let X be a divisor in $D_a(V \times W)$. Then there exist two divisors $Y \in D_a(V)$ and $Z \in D_a(W)$ such that $X \sim Y \times W + V \times Z$.*

Proof: We are going to use the theorem of the cube. By definition, we can write

$$X = T(u_0) - T(u_1)$$

with points u_0, u_1 on an abelian variety J, and a divisor T on the product $J \times V \times W$. Let k be a field of definition for J, V, W over which T is rational. We may suppose u_0 and u_1 generic over k (but not necessarily independent) after adding to them a generic point of J. The theorem of the cube asserts that $\sum(-1)^{i+j+k}T_{ijk} \sim 0$ on $J \times J \times V \times V \times W \times W$, the notations being those of Chapter III, § 2. Let v_1, w_1 be independent generic points of V, W over $k(u_0, u_1)$, and take the intersection of $\sum (-1)^{i+j+k} T_{ijk}$ with $u_0 \times u_1 \times V \times v_1 \times W \times w_1$. We then obtain the following cycles.

For $j = k = 1$, we find 0, since (u_i, v_1, w_1) cannot be in any component of T.

For $j = k = 0$, we obtain the two terms occurring in the expression for X, that is to say, if we project on $V \times W$ on the third and fifth factors, we find $T(u_0) - T(u_1)$.

The other terms give divisors of type $Y \times W$ and $V \times Z$. One sees directly from the definition of algebraic equivalence that $Y \in D_a(V)$ and $Z \in D_a(W)$.

REMARK. If V, W are complete and non-singular in codimension 1, then we can give an alternate proof for the preceding proposition, by taking $X(v)$ for v generic on V, and using the seesaw principle. We leave the details to the reader, observing merely that there are fewer indices in this procedure than in the one using the theorem of the cube.

PROPOSITION 7. *Let V, W be two varieties, complete and non-singular in codimension 1, both defined over k. Let (A, D) and (B, E) be Picard varieties of V and W respectively defined over k. Let F be the divisor on $V \times W \times A \times B$ obtained from the divisor*

$$D \times W \times B + V \times A \times E$$

by the transformation which permutes the two middle factors. Then $(A \times B, F)$ is a Picard variety of $V \times W$, defined over k.

Proof: According to Proposition 6, it is clear that condition (1) in the definition of a Picard variety is satisfied. We must therefore prove that the rationality condition (2) is also satisfied. Suppose that $X \in D_a(V \times W)$, and that X is rational over a field $K \supset k$. Let P, Q be two independent generic points of V over K. By Proposition 5 of the Appendix, § 2 the linear equivalence classes of $Z, X(P)$, and $X(Q)$ are equal, Z being the divisor of Proposition 6. There exist two divisors in this class, rational over $K(P)$ and $K(Q)$, respectively. Hence the point $z = \mathrm{Cl}\,(Z)$ associated with the linear equivalence class of Z on W in the Picard variety of W is rational over $K(P) \cap K(Q) = K$. Similarly, the point $y = \mathrm{Cl}\,(Y)$ is rational over K. If v, w are independent generic points of A, B over K then

$$X \sim [{}^tD(y + v) - {}^tD(v)] \times W + V \times [{}^tD(z + w) - {}^tD(w)]$$

and consequently

$$X \sim {}^tF(y + v, z + w) - {}^tF(v, w),$$

thereby proving our proposition.

Historical note:

Matsusaka was the first to construct a Picard variety of a variety V over the given field of definition of V [53], by constructing a total family of divisors on V, i.e., an algebraic family $\mathfrak{F}$ of divisors such that every divisor algebraically equivalent to 0 on V is linearly equivalent to $X_1 - X_2$ for $X_1, X_2 \in \mathfrak{F}$. This is in fact derived from the methods of the Italian school. It is projective, depending on the Chow families, and leads to a projective analysis of the Picard variety. This aspect of the question is far from having been cleared up, and we shall return to it below.

Chow's method is quite different, and is based on the universal mapping property [15]. Nevertheless, Chow obtains the covariant and contravariant mappings derived from the Albanese and Picard varieties, and proves the fundamental theorem that the Picard variety of V can be obtained as an inverse image of the Picard variety of its Albanese variety. He constructs the Picard variety of an abelian variety only at the end of his theory, whereas here, we construct it essentially at the beginning. His theory of the K/k-image and K/k-trace play an essential role, and allow him to obtain the duality. He needs a weak equivalence criterion since he obtains his Picard variety from the Jacobian of a generic curve.

One should also note here the articles of Morikawa [63] and Barsotti [5], which give a construction of the Picard variety of an abelian variety only up to a purely inseparable isomorphism, for lack of a method of taking quotients rationally.

On the other hand, in his treatise [85], Weil had already given a number of fundamental results on the divisor classes of an abelian variety for the various equivalences, and by means of these theorems, he recovers in the abstract case the classical equivalence criteria of Severi [89]. As we have already mentioned in the preceding chapter, he succeeded in proving the general formulation of the theorems of the square and the cube from intersection theory and the properties of the Jacobian. From there, he shows how the kernel of the homomorphism $\varphi_X : u \to \mathrm{Cl}\,(X_u - X)$ is an algebraic group, and how one can construct the Picard variety by finding non-degenerate divisors on A (this being also Morikawa's method [63]). Weil's proof turns out to be simple, because of the theorem of the square, and we have followed his construction (unpublished). One should also note that the definition of the Picard variety used here is due to him.

Of course, proving the existence of a non-degenerate divisor X such that φ_X maps A surjectively on Pic (A) amounts to constructing a total family, but it is parametrized immediately by an abelian variety. One avoids the use of the Chow families by means of Proposition 1 of Chapter III, § 1 which asserts that a cycle which is algebraically equivalent to 0 comes from an algebraic family parametrized by an abelian variety. This is an existence theorem, and the Jacobian plays a crucial role in its proof. It is again used in the theorem of the square, and thus in the proof of Theorem 4, due to Weil, and distilled from similar arguments in his treatise [85].

Weil also recognized the power of the theory of correspondences, which is used throughout through the seesaw principle (implicit in Severi's theory of correspondences). All the above methods are entirely birational or biholomorphic in character.

We return to projective theories with Weil's proof of the projective embedding [93], adapted from Lefschetz, which gives rise in a natural fashion to the notion of polarized abelian variety. The characterization of positive non-degenerate divisors in terms of hyperplane sections was given by Weil [93], and we have used here another method to prove one half of the theorem, better adapted to the means at our disposal.

The notion of a polarized abelian variety was isolated by Weil in the theory of complex multiplication [96] and in Torelli's theorem [94]. It gives a structure which is stronger than the biholomorphic structure, and its group of automorphisms is finite (a fact due to Matsusaka). We shall give Weil's proof in Chapter VIII.

Projective invariants are also essential in treating questions concerning algebraic families of abelian varieties. Without even speaking of the theory of moduli, one does not know the answer to the following question. Let A be an abelian variety and A' be a specialization of A (possibly a reduction mod p) such that A' has one component of multiplicity 1 which is a non-singular variety. Is A' an abelian variety, whose law of composition is obtained by specializing that of A? The only result in this direction is due to Deuring [30] and Chow-Lang [22], and gives the uniqueness of the non-degenerate specialization, if it is an abelian variety.

In this direction, one should determine over an algebraic number field whether the exceptional localities in reducing mod $\mathfrak{p}$ coincide with the exceptional localities in Taniyama's zeta function [81]. For elliptic curves, this is due to Deuring [30]. In higher dimension, one should also determine the conductor.

Of course, the general problem of algebraic systems of abelian varieties is the problem of moduli, and a first step in this direction has been taken by Matsusaka [61]. Here also, the problem is projective, and consists in classifying polarized abelian varieties.

Matsusaka's construction of the Picard variety is roughly analogous to Chow's construction of the Jacobian, and comes closest to giving the analogue to Igusa's compatibility theorem. Nevertheless, it falls considerably short of this goal, and further investigations in this direction are essential.

Once certain results concerning numerical equivalence have been obtained, Lang [48] has shown how one can recover and complete numerical theorems of Weil [85], and how one obtains canonical expressions for the trace of an endomorphism, using only the theorem of the square, and the fact that a divisor $X \equiv 0$ is numerically equivalent to 0. We have essentially copied [48] in this chapter and the next.

Finally, a propos of equivalence relations on an abelian variety, no progress has been made on the theory of intermediate cycles. For instance, the problem raised by Weil to determine whether every cycle Z is numerically equivalent to Z^- remains open. Actually, one is only beginning to have solid bases for equivalence theories on arbitrary varieties, and in particular, for linear equivalence (see Chow [19] and Samuel [75]). Presumably, the theory of intermediate cycles would show that the representation of an endomorphism on the cycles of dimension i is of degree $2r - 2i$ (with $r = \dim A$). This is trivial in the classical case, in view of the fact that the topological 1-cycles generate the homology ring.

CHAPTER V

Functorial Formulas

We first define the transpose of a homomorphism, i.e., the contravariant mapping induced on the Picard varieties. We prove that the transpose of an exact sequence (up to isogenies) is exact (up to isogenies).

Next, we have a large number of formal results relating to each other all sorts of mappings and induced mappings surrounding the Picard variety.

They are applied in the final section to the study of the trace of endomorphisms. If X is a positive non-degenerate divisor on an abelian variety, and $\varphi_X : A \to \hat{A}$ now denotes the isogeny on the Picard variety, then we can define an involution on the algebra of endomorphisms $\mathrm{End}_{\mathbf{Q}}(A)$, by putting

$$\alpha' = \alpha'_X = \varphi_X^{-1}\, {}^t\alpha\, \varphi_X ,$$

with $\alpha \in \mathrm{End}_{\mathbf{Q}}(A)$, and $\varphi_X^{-1} \in H_{\mathbf{Q}}(\hat{A}, A)$. The fundamental theorem on the algebra of endomorphisms then states that the bilinear form $(\alpha, \beta) \to \mathrm{tr}(\alpha'\beta)$ is positive definite, i.e., that $\mathrm{tr}(\alpha'\alpha) > 0$ for $\alpha \neq 0$. If the abelian variety A is defined over a finite field with q elements, and if π denotes the Frobenius endomorphism relative to k, then we recover immediately the theorem that all the characteristic roots of π have absolute value $q^{1/2}$: this is the Riemann hypothesis for abelian varieties.

§ 1. *The transpose of a homomorphism*

Let $\alpha : A \to B$ be a homomorphism of an abelian variety into another. Suppose that A, B are defined over k. Let $(\hat{A}, D)$ be a Picard variety of A, and $(\hat{B}, E)$ a Picard variety of B, both defined over k. We are going to see that α induces a homomorphism of $\hat{B}$ into $\hat{A}$. We need first an auxiliary result.

PROPOSITION 1. *Let $f : U \to V$ be a rational map of one variety into another, and assume that V is complete and non-singular. Let Y be a divisor in $D_a(V)$, such that $f^{-1}(Y)$ is defined. Then $f^{-1}(Y) \in D_a(U)$.*

Proof: There exists an abelian variety J and a divisor Z on $V \times J$ such that $Y \sim {}^tZ(x) - {}^tZ(y)$, with two points x, y of J (Proposition 1 of Chapter III, § 1). By the theorem of the square, we may assume that x, y are generic on J, adding to them if necessary a generic point of J. This does not change the linear equivalence class (Corollary 2 of Theorem 4, Chapter III, § 3). Let Γ be the graph of f. We can add to Z the divisor of a function on $V \times J$ to get a divisor Z_1 such that $(\Gamma \times J) \cdot (U \times Z_1)$ is defined (Proposition 4 of the Appendix, § 2). We still have $Y \sim {}^tZ_1(x) - {}^tZ_1(y)$ with x, y generic. We can form the composed divisor $T = \Gamma \circ Z_1$ as in Theorem 7 of the Appendix, § 2, and we see from this theorem that

$$f^{-1}(Y) \sim {}^tT(x) - {}^tT(y),$$

thus showing that $f^{-1}(Y)$ is algebraically equivalent to 0.

Let us take a linear equivalence class of $D_a(B)/D_l(B)$, and let Y be a representative divisor. As we have seen in Chapter III, § 1 we can define the inverse of this class to be a linear equivalence class of A, and our proposition shows that we get an element of Pic (A). Furthermore, if Y is rational over a field $K \supset k$, we can always change Y by a divisor linearly equivalent to 0 over K such that the inverse image is defined. (This has been pointed out in the Appendix, § 2.) If $\alpha^{-1}(Y)$ is defined, then it is also rational over K. It follows that the point on the Picard variety $\hat{A}$ of A associated with $\alpha^{-1}(Y)$ is rational over K. In this manner, we obtain a homomorphism

$${}^t\alpha : \hat{B} \to \hat{A}$$

defined by ${}^t\alpha(\mathrm{Cl}\,(Y)) = \mathrm{Cl}\,(\alpha^{-1}(Y))$ whenever $\alpha^{-1}(Y)$ is defined for $Y \in D_a(B)$. Here Cl denotes the point on the Picard variety, and we emphasize that ${}^t\alpha$ is now a rational homomorphism, and not merely a set-theoretic abstract homomorphism.

We have the following formulas:

t1. If $\alpha : A \to B$ and $\beta : B \to C$ are two homomorphisms then

$$^t(\beta\alpha) = {}^t\alpha{}^t\beta.$$

t2. If α, $\beta : A \to B$ are two homomorphisms, then

$$^t(\alpha + \beta) = {}^t\alpha + {}^t\beta.$$

t3. ${}^t\delta_A = \delta_A$.

The first formula comes from Theorem 2 of the Appendix, § 1. The second is a consequence of the corollary to Theorem 2, Chapter IV, § 1, and of Proposition 1. The last one is obvious.

PROPOSITION 2. *Let $\alpha : A \to B$ be a homomorphism. If α is surjective, then the kernel of ${}^t\alpha$ is finite. If the kernel of α is finite, then ${}^t\alpha$ is surjective; α is an isogeny if and only if ${}^t\alpha$ is an isogeny.*

Proof: Suppose first that α is an isogeny. We can find a homomorphism $\beta : B \to A$ such that $\beta\alpha = n\delta_A$ for some n (Chapter II, § 1 justified by Corollary 1 of Theorem 3, Chapter IV, § 2). Taking the transpose, we get ${}^t\alpha{}^t\beta = n\delta$. This shows that ${}^t\alpha$ is an isogeny. Suppose next that the kernel of α is finite. According to Poincaré's complete reducibility theorem (Chapter II, § 1 Theorem 6) there exists a homomorphism $\beta : B \to A$ such that $\beta\alpha = n\delta_A$. It follows again that ${}^t\alpha{}^t\beta = n\delta$, and hence that ${}^t\alpha$ is surjective. Finally, suppose that α is surjective. Again by Poincaré's theorem, we can find a homomorphism $\beta : B \to A$ such that $\alpha\beta = n\delta_B$, whence ${}^t\beta{}^t\alpha = n\delta$. The kernel of ${}^t\alpha$ must therefore be finite. This proves our proposition.

PROPOSITION 3. *Let $0 \to A \xrightarrow{\alpha} B \xrightarrow{\beta} C \to 0$ be an exact sequence, up to isogenies. Then the transposed sequence*

$$0 \leftarrow \hat{A} \xleftarrow{{}^t\alpha} \hat{B} \xleftarrow{{}^t\beta} \hat{C} \leftarrow 0$$

is exact, up to isogenies.

Proof: We have seen in Proposition 2 that the kernel of ${}^t\beta$ is finite and that ${}^t\alpha$ is surjective. Since we have $\beta\alpha = 0$, we have ${}^t\alpha{}^t\beta = 0$. This shows that the image of ${}^t\beta$ is contained in the kernel of ${}^t\alpha$. Since $\dim B = \dim A + \dim C$, we must have $\dim \hat{B} = \dim \hat{A} + \dim \hat{C}$. From these remarks, it follows immediately that the connected component of the kernel of ${}^t\alpha$ is equal to the image of ${}^t\beta$.

We shall give in a later chapter more precise results concerning the transpose of an exact sequence, in the case where α and β are regular.

§ 2. *A list of formulas and commutative diagrams*

Let T be a divisor on a product $A \times B$ of abelian varieties. Let k be a field of definition for A, B over which T is rational. We can then define homomorphisms (i.e., rational homomorphisms)

$$\lambda_T : A \to \hat{B}, \quad \lambda'_T : B \to \hat{A}$$

by the formulas

$$\lambda_T(u) = \mathrm{Cl}[T(u+v) - T(v)], \quad \lambda'_T(w) = \mathrm{Cl}[{}^tT(w+z) - {}^tT(z)],$$

taking u, v generic independent on A and w, z generic independent on B. Suppose that $\lambda_T = 0$. This means that $T(u+v) \sim T(v)$ for all u, v, or in other words that the class $T(v)$ is constant. The seesaw principle (Appendix, Theorem 6) shows that $\lambda'_T = 0$. By symmetry, we have $\lambda_T = 0$ if and only if $\lambda'_T = 0$. Furthermore, the composition of correspondences as described in Theorem 7 of the Appendix, § 2 gives us the following theorem.

PROPOSITION 4. *Let T be a divisor on a product $A \times B$. Then the map*

$$T \to \lambda_T$$

induces an isomorphism of the correspondence classes on $A \times B$ into $H(A, \hat{B})$. If A, B are defined over k, and if T is rational over k, then λ_T is defined over k. The above isomorphism is actually surjective; for, if $\lambda \in H(A, \hat{B})$ is given and defined over k, then there exists a Poincaré divisor E on $B \times \hat{B}$ rational over k such that the composed correspondence ${}^tE \circ \lambda$ is defined, and if we put $T = {}^tE \circ \lambda$, then $\lambda = \lambda_T$.

Proof: The fact that $T \to \lambda_T$ induces an isomorphism of the correspondence classes of $A \times B$ into $H(A, \hat{B})$ has already been seen above. Let $\lambda : A \to \hat{B}$ be a homomorphism defined over k. As we have seen many times, we can change a given Poincaré divisor by a linear equivalence defined over k, in order to obtain such a divisor E for which ${}^tE \circ \lambda$ is defined. Our assertion is then a consequence of the definition of the Poincaré divisor.

We see that if $\lambda : A \to \hat{B}$ is a homomorphism, the divisor T such that $\lambda = \lambda_T$ is uniquely determined up to a trivial divisor. If $\lambda, \tau : A \to \hat{B}$ are two homomorphisms, we have $\lambda = \tau$ if and only $\lambda' = \tau'$. The mapping

$$\lambda \to \lambda'$$

establishes an isomorphism between $H(A, \hat{B})$ and $H(B, \hat{A})$.

In Proposition 6 below we shall give another way of finding a divisor T associated with λ. Before doing this, we need to consider a special case of the homomorphism λ_T. Let us take $B = \hat{A}$ and let T be a Poincaré divisor on $A \times \hat{A}$. We know that this divisor is uniquely determined up to a trivial one, and thus we obtain a *canonical homomorphism*

$$\kappa_A : A \to \hat{\hat{A}}$$

of A into the Picard variety of $\hat{A}$. We sometimes call the Picard variety of A its *dual* variety, and thus the preceding homomorphism is into the double dual. (It will be easy to see below that κ_A is an isogeny.) We have by definition $\kappa_A = \lambda_D$ where D is a Poincaré divisor on $A \times \hat{A}$, and

$$\lambda'_D = \delta_{\hat{A}}$$

by definition. (See p. 229.)

PROPOSITION 5. *Let T be a divisor on $A \times B$, and $\lambda = \lambda_T$. Then the following diagram is commutative:*

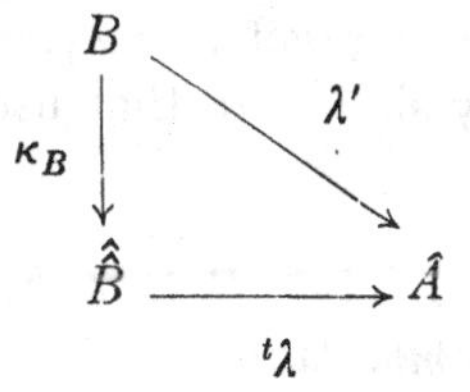

Proof: Proposition 4 shows that we can take for T the divisor $F = {}^tE \circ \lambda$, with a Poincaré divisor E on $B \times \hat{B}$. We have ${}^tF = {}^t\lambda \circ E$, that is to say, for two independent generic points

$$A \xrightarrow{\lambda} \hat{B} \overset{{}^tE}{\times} B$$

v, w of B, we have

$${}^tF(v + w) - {}^tF(w) \sim \lambda^{-1}[E(v + w) - E(w)].$$

From the definitions, we get $\lambda' = {}^t\lambda\kappa_B$.

PROPOSITION 6. *Let* $\lambda : A \to \hat{B}$ *be a homomorphism, and let* $(\lambda, \delta) : A \times B \to \hat{B} \times B$ *be the product of* λ *with the identity. Let* E *be a Poincaré divisor on* $B \times \hat{B}$ *such that the divisor* $T = (\lambda, \delta)^{-1}({}^tE)$ *is defined. Then we have* $\lambda = \lambda_T$.

Proof: Let v be a generic point of B, and consider the following commutative diagram:

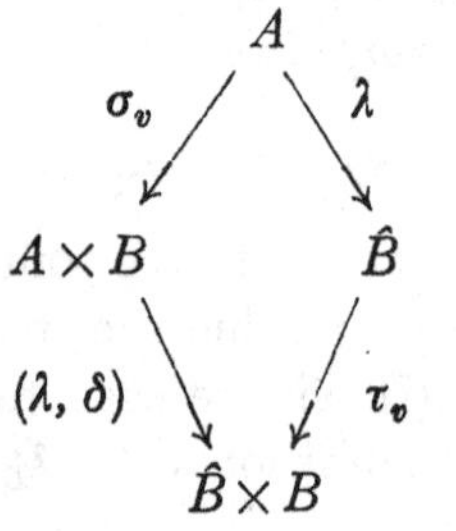

where σ_v is such that $\sigma_v(u) = (u, v)$ and τ_v is such that $\tau_v(\hat{w}) = (\hat{w}, v)$. If we take the inverse image of ${}^tE$ on the right, we find $\lambda^{-1}(E(v))$. Going around on the left, and using Theorem 2 of the Appendix, § 1 we obtain ${}^t\lambda\kappa_B = \lambda'_T$. Proposition 5 shows that if W is a divisor on $A \times B$ such that $\lambda = \lambda_W$, we have $\lambda'_W = \lambda'_T$. By the symmetry of the seesaw principle, we get $\lambda = \lambda_T$.

In a manner entirely similar to that used in the proof of Proposition 5, we get

PROPOSITION 7. *Let* $\alpha : A \to B$ *be a homomorphism. Then the following diagram is commutative:*

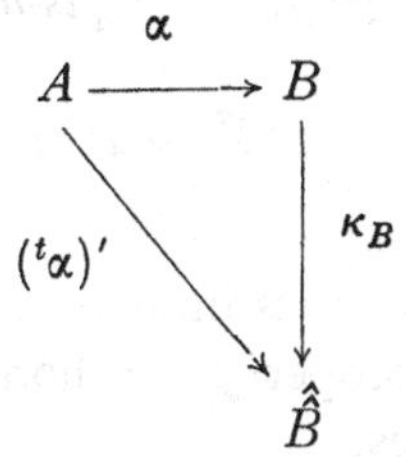

Combining Propositions 5 and 7, we find for $\lambda = {}^t\alpha$:

PROPOSITION 8. *Let* α: $A \to B$ *be a homomorphism. Then the following diagram is commutative*:

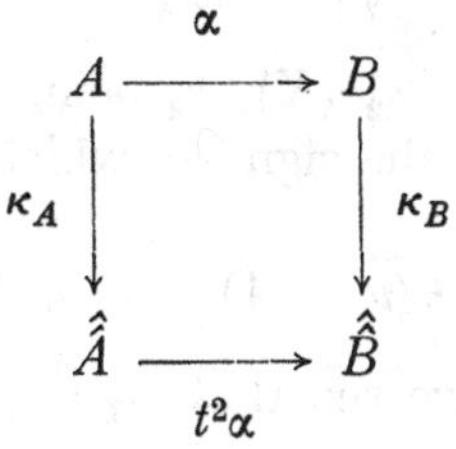

Of course, by $t^2\alpha$ we mean ${}^t({}^t\alpha)$.

We shall say that an abelian variety is *reflexive* if κ_A is a birational isomorphism. We can then use κ_A to identify A and $\hat{\hat{A}}$.

Let us take $\lambda = \kappa_A$, and apply Proposition 5. We have already observed that $\lambda' = \delta_{\hat{A}}$. The diagram in Proposition 5 becomes

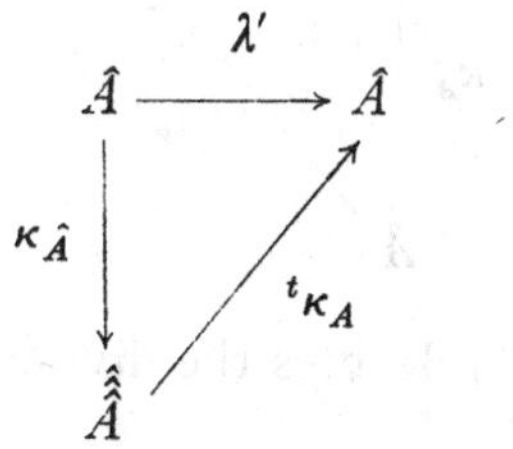

Since we have $1 = \nu(\delta_{\hat{A}}) = \nu(\lambda') = \nu(\kappa_A)\nu({}^t\kappa_A)$, we obtain the following result.

PROPOSITION 9. *For every abelian variety* A, *the canonical*

homomorphism κ_A is an isogeny, and ${}^t\kappa_A$ is a birational isomorphism. So is

$$\kappa_{\hat{A}} : \hat{A} \to \hat{\hat{\hat{A}}}$$

and thus $\hat{A}$ is reflexive.

Using Proposition 9, one sees immediately that $(\hat{\hat{A}}, \kappa_A)$ satisfies the universal mapping property for homomorphisms of A into reflexive abelian varieties.

Let X be a divisor on A. We had defined $\varphi_X : A \to \mathrm{Pic}\,(A)$ by $\varphi_X(u) = \mathrm{Cl}\,(X_u - X)$. We can now take φ_X as a rational homomorphism of A into $\hat{A}$, defined by the same formula. One sees immediately that

$$\varphi_X = \lambda_T,$$

where T is the divisor $-s_2^{-1}(X)$, s_2 being as usual the sum on $A \times A$. Note especially the sign $-$, which comes from the fact that

$$s_2^{-1}(X) \cdot (u \times A) = u \times X_{-u}.$$

Since T is symmetric, we see that $\varphi_X = \varphi'_X$, and Proposition 5 becomes

PROPOSITION 10. *Let X be a divisor on A. Then $\varphi_X = \varphi'_X$ and the following diagram is commutative:*

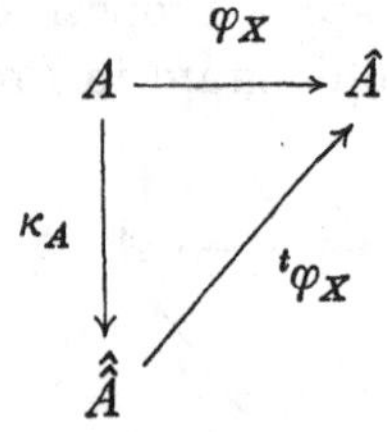

We recall that $N'(A)$ denotes the divisor classes for the square equivalence $\equiv$.

PROPOSITION 11. *Let $\alpha: A \to B$ be a homomorphism, and let $\eta \in N'(B)$. Let ξ be the element $\alpha^{-1}(\eta)$ in $N'(A)$. Then the following diagram is commutative:*

$$\begin{array}{ccc} A & \xrightarrow{\alpha} & B \\ {\scriptstyle \varphi_\xi}\downarrow & & \downarrow{\scriptstyle \varphi_\eta} \\ \hat{A} & \xleftarrow[{}^t\alpha]{} & \hat{B} \end{array}$$

Proof: This is an immediate consequence of Lemma 1, Chapter IV, § 1.

Before stating the next proposition, we recall that if $\varphi : A \times B \to A_1 \times B_1$ is a homomorphism of a product of abelian varieties into another product, then we can represent φ by a matrix,

$$\varphi = \begin{pmatrix} \alpha & \beta \\ \alpha_1 & \beta_1 \end{pmatrix}$$

with suitable homomorphisms α, β, α_1, β_1 such that for (u, v) on $A \times B$ we have $\varphi(u, v) = (\alpha u + \beta v, \alpha_1 u + \beta_1 v)$. Thus we consider (u, v) as a vertical vector on which our matrix operates.

PROPOSITION 12. *Let T be a divisor on a product $A \times B$, so that $\varphi_T : A \times B \to \hat{A} \times \hat{B}$ is a homomorphism. Then*

$$\varphi_T = \begin{pmatrix} \alpha & -\lambda'_T \\ -\lambda_T & \beta \end{pmatrix}$$

with two suitable homomorphisms α, β, which we do not determine explicitly.

Proof: We can write

$$\varphi_T = \begin{pmatrix} \alpha & \sigma \\ \tau & \beta \end{pmatrix}.$$

By definition, we have

$$(\alpha u, \tau u) = \mathrm{Cl}\,[T_{(u, 0)} - T].$$

On the other hand, we have $\lambda_T u = \mathrm{Cl}[T(u) - T(0)]$ (we may assume that $T(0)$ is defined, replacing T by a divisor linearly equivalent to it if necessary). Making a translation to the intersection

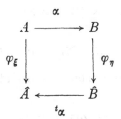

$$T \cdot (u \times B) = u \times T(u)$$

we find

$$T_{(-u,0)} \cdot (0 \times B) = 0 \times T(u).$$

Let D, E be Poincaré divisors on $A \times \hat{A}$ and $B \times \hat{B}$, respectively. By definition, we have

$$T_{(u,0)} - T \sim [{}^tD(\alpha u) - {}^tD(0)] \times B + A \times [{}^tE(\tau u) - {}^tE(0)].$$

This gives

$$T(-u) - T(0) \sim {}^tD(\tau u) - {}^tD(0).$$

In view of our preceding remarks, we get $\lambda_T = -\tau$. By symmetry, we must then have $\sigma = -\lambda'_T$. This proves our proposition.

§ 3. *The involutions*

Let A, B be two abelian varieties, and let $r = \dim A$.

We are going to study more closely the divisor of Proposition 2 of Chapter IV, § 1. We work again with the square equivalence $\equiv$, and as in Chapter IV, we denote by $N'(A)$ the factor group of the divisors on A by the subgroup of divisors X such that $\varphi_X = 0$. We recall that if $\eta \in N'(B)$, then for α, $\beta \in H(A, B)$ we put

$$D_\eta(\alpha, \beta) = (\alpha + \beta)^{-1}(\eta) - \alpha^{-1}(\eta) - \beta^{-1}(\eta).$$

It is an element of $N'(A)$. In view of Proposition 1 of § 1, and of the definitions, we find by an argument similar to the one used in Proposition 2 of Chapter IV, § 1 that

$$\varphi_{D_\eta(\alpha,\beta)} = {}^t\alpha\varphi_\eta\beta + {}^t\beta\varphi_\eta\alpha. \tag{1}$$

The linearity of ${}^t\alpha$ and the symmetry of this formula shows us that the symbol $D_\eta(\alpha, \beta)$ satisfies the following conditions:

D1. $D_\eta(\alpha + \beta, \gamma) = D_\eta(\alpha, \gamma) + D_\eta(\beta, \gamma)$,

D2. $D_\eta(\alpha, \beta) = D_\eta(\beta, \alpha)$,

D3. $D_\eta(m\alpha, \beta) = m \cdot D_\eta(\alpha, \beta)$,

D4. $D_\eta(\alpha, \alpha) = 2 \cdot \alpha^{-1}(\eta)$.

This last formula is due to the commutative diagram of § 2,

Proposition 11. We see that the mapping

$$(\alpha, \beta) \to D_\eta(\alpha, \beta)$$

is a bilinear form on $H(A, B) \times H(A, B)$ taking its values in $N'(A)$. We can obviously extend it to the tensor products with the rationals, i.e., define $D_\eta(\alpha, \beta)$ for $\alpha, \beta \in H_{\mathbf{Q}}(A, B)$ and $\eta \in N'_{\mathbf{Q}}(B)$. We suppose in the sequel that this extension has been made.

As in Chapter IV, § 3 we put

$$D_\xi(\alpha) = D_\xi(\alpha, \delta)$$

in case $A = B$ and $\xi = \eta$ is an element of $N'_{\mathbf{Q}}(A)$. We then see that the map

$$\alpha \to D_\xi(\alpha)$$

is a linear map of $\mathrm{End}_{\mathbf{Q}}(A)$ into $N'_{\mathbf{Q}}(A)$.

For a composition of mappings, we immediately get from (1) and the commutative diagram of Proposition 11 of § 2 the following formulas:

D5. $D_\zeta(\lambda\alpha, \lambda\beta) = D_{\lambda^{-1}(\zeta)}(\alpha, \beta)$,

D6. $D_\eta(\alpha\lambda, \beta\lambda) = \lambda^{-1} D_\eta(\alpha, \beta)$.

Of course, the composed maps must make sense. In D5, we must have $\lambda : B \to C$ and in D6 we have $\lambda : C \to A$. Furthermore, in D5, $\zeta \in N'(C)$, while in D6, $\eta \in N'(B)$.

Let ξ be an element of $N'_{\mathbf{Q}}(A)$ which is non-degenerate. By this we mean that some integral multiple of it contains a non-degenerate divisor. This is equivalent to saying that there exists an inverse φ_ξ^{-1} in $H_{\mathbf{Q}}(\hat{A}, A)$. Let $\alpha : A \to B$ be a homomorphism, and $\eta \in N'(B)$. We define

$$t_\xi{}^\eta(\alpha) = \varphi_\xi^{-1}\, {}^t\alpha\, \varphi_\eta\,.$$

It is essential here that ξ is non-degenerate. We may of course extend this to elements $\alpha \in H_{\mathbf{Q}}(A, B)$ and $\eta \in N_{\mathbf{Q}}(B)$, since the transpose of a homomorphism can be extended to $H'_{\mathbf{Q}}(A, B)$ by linearity. If $\lambda : B \to C$ is a homomorphism, or is an element of $H_{\mathbf{Q}}(B, C)$, and $\zeta \in N'_{\mathbf{Q}}(C)$, then the following formula is obvious from the properties of the transpose:

$$t_\xi{}^\eta(\alpha)\, t_\eta{}^\zeta(\lambda) = t_\xi{}^\zeta(\lambda\alpha). \tag{2}$$

We must of course assume that η is non-degenerate.

We also have an important formula which will be useful in proving the equivalence between numerical equivalence and the equivalence of the square. We take two elements α, β of $H_{\mathbf{Q}}(A, B)$, and $\eta \in N'_{\mathbf{Q}}(B)$, $\xi \in N'_{\mathbf{Q}}(A)$ with ξ non-degenerate,

D7. $D_\xi(\varphi_\xi^{-1}\,{}^t\alpha\,\varphi_\eta \beta) = D_\eta(\alpha, \beta)$.

To prove it, we merely use the commutative diagrams of Propositions 8 and 10 of § 2, taking (1) into account.

We can take $A = B$, and we can define an involution on $\operatorname{End}_{\mathbf{Q}}(A)$, taking $\eta = \xi$ in the definition of $t_\xi^\eta(\alpha)$. We put

$$\alpha'_\xi = \varphi_\xi^{-1}\,{}^t\alpha\,\varphi_\xi,$$

or simply α' if the reference to ξ is determined once and for all. The mapping $\alpha \to \alpha'$ is an anti-automorphism of $\operatorname{End}_{\mathbf{Q}}(A)$, that is to say, we have

$$(\alpha + \beta)' = \alpha' + \beta', \quad (\alpha')' = \alpha, \quad (\alpha\beta)' = \beta'\alpha', \quad \delta' = \delta.$$

The second equation comes from Propositions 10 and 8 of § 2. The others are obvious.

We may rewrite (1) in terms of our involution,

$$\varphi_{D_\xi(\alpha,\beta)} = \varphi_\xi(\alpha'\beta + \beta'\alpha), \tag{3}$$

and we can supplement the preceding formulas by others which describe the behavior of our symbol $D_\xi(\alpha, \beta)$ under multiplication, α and β being now elements of $\operatorname{End}_{\mathbf{Q}}(A)$:

D8. $D_\xi(\alpha, \beta) = D_\xi(\alpha'\beta)$ and $D_\xi(\alpha) = D_\xi(\alpha')$

We get D8 by putting $\xi = \eta$ in D7. If we apply it to $D_\xi(\alpha\beta, \gamma)$, we find $D_\xi(\beta'\alpha'\gamma)$. Going backwards, and leaving $\alpha'\gamma$ together, we find

D9. $D_\xi(\alpha\beta, \gamma) = D_\xi(\beta, \alpha'\gamma)$.

In other words, with respect to left multiplication, $\operatorname{End}_{\mathbf{Q}}(A)$ is a sort of Hilbert space, and our involution gives the transposition. In particular, we have

D10. $D_\xi(\beta'\alpha\beta) = \beta^{-1}D_\xi(\alpha)$.

Let us return to numerical equivalence. As we have already seen in the proofs of Theorems 6 and 7, of Chapter IV, § 3, a relation in $N'_{\mathbf{Q}}(A)$ gives rise to a relation for numerical equivalence, in $N_{*\mathbf{Q}}(A)$. We use especially D8. If we take now for ξ a polar class, i.e., an element of $N_{r-1}(A) \otimes \mathbf{Q}$ such that a suitable positive multiple contains a polar divisor, or an ample one, we can define a bilinear form

$$(\alpha, \beta) = \operatorname{tr}(\alpha'\beta)$$

on $\operatorname{End}_{\mathbf{Q}}(A)$, and we then have the fundamental theorem on the algebra of endomorphisms of an abelian variety.

THEOREM 1. *Let $\alpha, \beta \in \operatorname{End}_{\mathbf{Q}}(A)$, and let $\xi \in N_{r-1}(A)$ be such that a positive multiple $m \cdot \xi$ $(m > 0)$ contains an ample divisor. Let $\alpha' = \alpha'_{\xi}$. Then*

$$(\alpha, \beta) = \operatorname{tr}(\alpha'\beta)$$

is a positive definite bilinear form, i.e., is such that $(\alpha, \alpha) > 0$ if $\alpha \neq 0$, and we have

$$\operatorname{tr}(\alpha'\beta) = \frac{r}{\deg(\xi^r)} \deg(\xi^{r-1} \cdot D_{\xi}(\alpha, \beta)).$$

Proof: Since $D_{\xi}(\alpha, \alpha) = 2 \cdot \alpha^{-1}(\xi)$, the hypothesis on ξ implies that $m \cdot \alpha^{-1}(\xi)$ contains a positive divisor if $\alpha \neq 0$, and hence that the intersection which gives $\operatorname{tr}(\alpha'\alpha)$ in the theorem is positive.

Properties D2, D9, and D8 give us analogous properties for the trace, namely

tr1. $(\alpha, \beta) = (\beta, \alpha)$,

tr2. $(\alpha\beta, \gamma) = (\beta, \alpha'\gamma)$,

tr3. $\operatorname{tr}(\alpha) = \operatorname{tr}(\alpha')$.

In addition, we obtain a corollary which completes our result on numerical equivalence.

COROLLARY. *Let Y be a divisor on A such that Y is numerically equivalent to 0. Then $Y \equiv 0$.*

Proof: We use D7. Let η be the class of Y for numerical equivalence. For any endomorphism λ of A, we have $\lambda^{-1}(\eta) = 0$ for numerical equivalence. Consequently, for $\alpha, \beta \in \operatorname{End}(A)$, we have $D_\eta(\alpha, \beta) \cdot \zeta = 0$ for all $\zeta \in N_1(A)$. In particular, $D_\eta(\alpha, \beta) \cdot \xi^{r-1} = 0$ and for any α, β we get by D7 and the expression for the trace in Theorem 7 of Chapter IV, § 3

$$\operatorname{tr}(\varphi_\xi^{-1}\, {}^t\alpha \varphi_Y \beta) = 0.$$

Take $\alpha = \delta$ and $\beta = (\varphi_\xi^{-1}\varphi_Y)'$. The trace can be equal to 0 only if $\varphi_\xi^{-1}\varphi_Y = 0$, and since ξ is non-degenerate, we must have $\varphi_Y = 0$, and thus $Y \equiv 0$.

As usual, we can define the norm $\|\alpha\| = (\alpha, \alpha)^{1/2}$ associated with our quadratic form. We are going to compute it for some special endomorphisms.

In the first place, for the identity, we have

$$\|\delta\| = \operatorname{tr}(\delta'\delta)^{1/2} = (\operatorname{tr}(\delta))^{1/2} = (2r)^{1/2}.$$

Let A be defined over a finite field with q elements, and let $\pi : A \to A$ be the Frobenius endomorphism, $\pi(u) = u^{(q)}$ which transforms u into its image under the Frobenius automorphism of the universal domain. Let X be a non-degenerate divisor on A, rational over k. We are going to show that if $\pi' = \pi'_X = \varphi_X^{-1}\, {}^t\pi\varphi_X$, then $\pi'\pi = \pi\pi' = q\delta$. By definition, we have

$$\pi'\pi = \varphi_X^{-1}\, {}^t\pi\varphi_X\pi.$$

Since X is rational over k, φ_X is defined over k, and hence

$$\varphi_X\pi = \pi_{\hat{A}}\, \varphi_X$$

where $\pi_{\hat{A}}$ is the Frobenius endomorphism of $\hat{A}$. It will suffice to prove that

$${}^t\pi\pi_{\hat{A}} = q\delta_{\hat{A}}.$$

By definition, for every point $y \in \hat{A}$, we have $\pi_{\hat{A}}(y) = y^{(q)}$. The set-theoretic inverse image $\pi^{-1}(Y)$ for a divisor Y of A consists of those points $a \in A$ such that $a^{(q)} \in \operatorname{supp}(Y)$. If y is the point associated with the divisor $Y \in D_a(A)$ in $\hat{A}$, then $y^{(q)}$ is associated with $Y^{(q)}$. In order to find ${}^t\pi(y)$, we must therefore consider the divisor $\pi^{-1}(Y^{(q)})$ and determine the multiplicities occurring in it. Set-theoretically, we find Y. On the other hand we have the following lemma.

LEMMA 1. *Let k be a perfect field, K a finitely generated regular extension of k of transcendence degree r, and q a power of the characteristic. Then $[K : K^q] = q^r$.*

Proof: Let $x_1, \ldots, x_r$ be a separating transcendence base. Then if we put $K_0 = k(x_1, \ldots, x_r)$, it is obvious that $[K_0 : K_0^q] = q^r$.

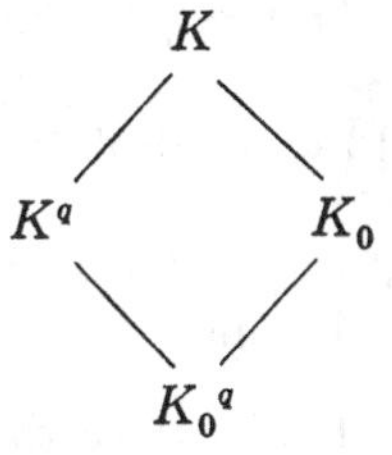

The extension K^q of K_0^q is isomorphic to K over K_0 and hence is separable. Hence K_0 and K^q are linearly disjoint over K_0^q. From all this, one sees immediately that $K = K_0 K^q$, and this proves our lemma.

If $r = \dim A$, we see that $\nu(\pi) = q^r$. According to Theorem 5 in the Appendix, we conclude that

$$\pi\pi^{-1}(Y^{(q)}) = q^r Y^{(q)}.$$

Let V be a subvariety of codimension 1 on A, and Γ_π the graph of π. Then

$$\pi(V) = \mathrm{pr}_2[\Gamma_\pi \cdot (V \times A)] = q^{r-1} V^{(q)},$$

as one sees immediately by applying the lemma to the function field of V. Since $\Gamma_\pi \cdot (V \times A)$ contains one component with multiplicity 1 (F—VII$_6$ Th. 17), we get the cycle $q^{r-1} V^{(q)}$ from the definition of pr. This shows that if we put $Z = \pi^{-1}(Y^{(q)})$ we have $\pi(Z) = q^{r-1} Z^{(q)}$. Hence $Z = qY$, as was to be shown.

Our computation implies that

$$\|\pi\|^2 = \mathrm{tr}(\pi'\pi) = \mathrm{tr}(q\delta) = 2rq.$$

On the other hand, $\mathrm{tr}(\pi) = \mathrm{tr}(\pi') = \mathrm{tr}(\pi'\delta) = (\pi, \delta)$ and the Schwartz inequality gives

$$|\mathrm{tr}(\pi)| \leq \|\pi\| \; \|\delta\| = 2rq^{1/2}.$$

We have seen in Chapter IV, § 3 that if ω_j are the characteristic roots of π then $\omega_j{}^n$ are the characteristic roots of π^n, and from the definition of the trace we get

$$\operatorname{tr}(\pi^n) = \sum_{j=1}^{2r} \omega_j{}^n\,, \qquad |\operatorname{tr}(\pi^n)| \leqq 2rq^{n/2}.$$

As we have seen above, $\nu(\pi^n) = q^{rn}$. This shows that

$$\prod_{j=1}^{2r} \omega_j = \pm q^r.$$

In order to show that $|\omega_j| = q^{1/2}$, it will therefore suffice to prove the following elementary lemma.

LEMMA 2. *Let ω_j $(j = 1, \ldots, s)$ be complex numbers, and z a real number such that*

$$\Big|\sum_{j=1}^{s} \omega_j{}^n\Big| \leqq sz^n$$

for all positive integers n. Then $|\omega_j| \leqq z$ for $j = 1, \ldots, s$.

Proof: If we bunch together the ω_j having the same absolute value, we see that it suffices to prove our lemma for each such bunch. Let w be this absolute value. We can write

$$\omega_j = we^{i\theta_j}$$

where θ_j is a real number. We have

$$\omega_j{}^n = w \sum_{j=1}^{s} e^{i\theta_j n}$$

and it will suffice to show that $\sum_{j=1}^{s} e^{i\theta_j n}$ comes arbitrarily close to s for infinitely many positive integers n. For this, we must show that the vector $(n\theta_1, \ldots, n\theta_s)$ comes arbitrarily close to $(0, \ldots, 0)$ for infinitely many n, modulo the vector periods of the exponential function. But we can partition the cube of dimension s into arbitrarily small cubes. For infinitely many n, the vectors $(n\theta_1, \ldots, n\theta_s)$ will yield distinct vectors. Their differences will therefore be arbitrarily close to the origin, modulo the periods. This proves the lemma.

The preceding results can be summarized as follows.

THEOREM 2. *Let A^r be an abelian variety defined over a field k with q elements, and let $\pi : A \to A^{(q)} = A$ be the Frobenius endomorphism. Then $\nu(\pi) = q^r$, and all the characteristic roots of π have absolute value $q^{1/2}$. If X is a polar divisor rational over k, and if $\pi' = \pi'_X$, then $\pi\pi' = \pi'\pi = q\delta$, and $\|\pi\| = \|\pi'\| = 2rq^{1/2}$.*

We also observe that our arguments in Lemma 2 show that $\mathrm{tr}(\pi^n)$ comes arbitrarily close to $2rq^{n/2}$ for infinitely many values of n.

Let V be a non-singular variety defined over a finite field k with q elements. Let $\mathfrak{p}$ range over all the prime rational cycles of dimension 0 of V over k. If s denotes a complex variable, the *zeta function* of V is defined by the infinite product

$$\zeta(s) = \prod_{\mathfrak{p}} \frac{1}{1 - \dfrac{1}{N\mathfrak{p}^s}}$$

where $N\mathfrak{p} = q^{\deg(\mathfrak{p})}$. It is convenient to change variables, and to write $t = q^{-s}$. Then the zeta function becomes

$$Z(t) = \prod_{\mathfrak{p}} \frac{1}{1 - t^{\deg(\mathfrak{p})}}$$

Taking the logarithmic derivative formally, we get

$$\frac{d}{dt} \log Z(t) = \sum_{n=1}^{\infty} N_n t^{n-1}$$

where N_n denotes the number of rational points of V over the unique extension k_n of degree n of k. The logarithmic derivative determines the zeta function uniquely provided that we normalize it to take the value 1 at $t = 0$. In all the cases with which we deal, we shall prove that the series for the logarithmic derivative converges in the circle $|t| < q^{-r}$ with $r = \dim V$, and thus that the infinite product converges in that circle.

It is trivial to see that the endomorphism $\pi^n - \delta$ of the abelian variety A defined over k is separable, and its kernel consists of the points $a \in A$ such that $a^{(q^n)} = a$. These are precisely the rational points of A over k_n. The number N_n is therefore given by

$\nu(\pi^n - \delta)$ which we can determine in terms of the characteristic roots ω_j of π by means of Theorem 8 of Chapter IV, § 3. We have

$$N_n = \prod_j (\omega_j{}^n - 1).$$

From this one verifies immediately for an abelian variety Weil's conjecture concerning the structure of the zeta function of a complete non-singular variety. This conjecture states that $Z(t)$ can be written as an alternating product

$$Z(t) = \frac{F_1(t) \ldots F_{2r-1}(t)}{F_0(t)F_2(t) \ldots F_{2r}(t)}$$

where $F_0(t) = 1 - t$, $F_{2r}(t) = 1 - q^r t$, and the $F_i(t)$ are polynomials in t with integer coefficients, of type

$$F_i(t) = \prod (1 - \rho_{i\mu} t)$$

where the roots $\rho_{i\mu}$ of $F_i(t)$ have absolute value $q^{i/2}$. If the variety is obtained by a non-degenerate reduction of a variety defined over the rational numbers (or an algebraic number field) then the degree of the polynomial $F_i(t)$ should be equal to the ith Betti number of this variety, viewed as a topological manifold. In the case of an abelian variety, which is a product of $2r$ circles, this is obvious.

In the next chapter, we shall determine the zeta function of a complete non-singular curve, and show that the absolute value of the roots of the polynomial in its numerator are equal to $q^{1/2}$. In fact, we shall prove that this polynomial is essentially the characteristic polynomial of the Frobenius endomorphism of its Jacobian. Thus if one can prove the Riemann hypothesis for curves, one can then derive it immediately for abelian varieties, taking into account the fact that an abelian variety is a homomorphic image of the Jacobian of a generating curve, together with Poincaré's complete reducibility theorem. Under our present approach to these questions, it has been more natural to derive first the theory on abelian varieties.

We conclude this chapter by some remarks on the positivity of the bilinear form of Theorem 1. We assume till the end that

ξ is a polar class in $N_{r-1}(A)$. We write $N_{\mathbf{Q}}(A)$ instead of $N_{r-1}(A) \otimes \mathbf{Q}$.

We shall say that an element α of $\mathrm{End}_{\mathbf{Q}}(A)$ is *symmetric* (relative to the involution induced by ξ) if $\alpha = \alpha'$. The symmetric elements form a subspace $S_{\mathbf{Q}}(A)$ of $\mathrm{End}_{\mathbf{Q}}(A)$ over the rationals. We have a $\mathbf{Q}$-linear map

$$D_\xi : S_{\mathbf{Q}}(A) \to N_{\mathbf{Q}}(A)$$

defined by $\alpha \to D_\xi(\alpha)$. It is in fact an isomorphism, and we can define its inverse

$$\Phi_\xi : N_{\mathbf{Q}}(A) \to S_{\mathbf{Q}}(A)$$

as follows: Let $\eta \in N_{\mathbf{Q}}(A)$. Put $\Phi_\xi(\eta) = \frac{1}{2}\varphi_\xi^{-1}\varphi_\eta$. The map $\eta \to \Phi_\xi(\eta)$ is obviously a $\mathbf{Q}$-linear map of $N_{\mathbf{Q}}(A)$ into $\mathrm{End}_{\mathbf{Q}}(A)$. The commutativity of Proposition 10 of § 2 shows that $\varphi_\xi^{-1}\varphi_\eta$ is symmetric so that Φ_ξ maps $N_{\mathbf{Q}}(A)$ into $S_{\mathbf{Q}}(A)$. Formula D7 shows that $D_\xi(\varphi_\xi^{-1}\varphi_\eta) = 2 \cdot \eta$ and (1) shows that $\varphi_\xi^{-1}\varphi_{D_\xi(\alpha)} = 2 \cdot \alpha$ if α is symmetric. Our mappings are therefore inverse to each other.

We agree to write

$$\alpha \longleftrightarrow \eta$$

if $\alpha = \Phi_\xi(\eta)$. This being the case, we have the following proposition, which is nothing else than a reformulation of D10.

PROPOSITION 13. *Let α be a symmetric element of* $\mathrm{End}_{\mathbf{Q}}(A)$, *and let $\alpha \longleftrightarrow \eta$. Let λ be any element of* $\mathrm{End}_{\mathbf{Q}}(A)$. *Then*

$$\lambda'\alpha\lambda \longleftrightarrow \lambda^{-1}(\eta).$$

In particular, $\lambda'\lambda \longleftrightarrow \lambda^{-1}(\xi)$.

We shall say that an element α of $\mathrm{End}_{\mathbf{Q}}(A)$ is *positive* (relative to the involution induced by ξ) if it is symmetric and if there exists an integer $m > 0$ such that $m \cdot D_\xi(\alpha)$ in $N_{r-1}(A)$ contains a positive divisor. We then write $\alpha > 0$.

PROPOSITION 14. *Let α be a symmetric element of* $\mathrm{End}_{\mathbf{Q}}(A)$. *Then*:

(i) *If $\alpha \geqq 0$ and β is an arbitrary element of* $\mathrm{End}_{\mathbf{Q}}(A)$ *we have* $\beta'\alpha\beta \geqq 0$.

(ii) *If* $\beta \in \mathrm{End}_{\mathbf{Q}}(A)$ *and* $\beta \neq 0$, *then* $\beta'\beta > 0$.

(iii) *If* $\alpha > 0$ *then* $\mathrm{tr}(\alpha) > 0$.

Proof: These assertions are obvious from the results already obtained.

Further properties of the trace can be axiomatized.

THEOREM 3. *Let R be a finite dimensional algebra over the rationals $\mathbf{Q}$, with an involution $x \to x'$, i.e., a $\mathbf{Q}$-linear map onto itself such that $(xy)' = y'x'$ and $(x')' = x$. Assume that there exists a $\mathbf{Q}$-linear functional $\sigma : R \to \mathbf{Q}$ satisfying $\sigma(xy) = \sigma(yx)$, such that for $x \neq 0$ we have $\sigma(xx') > 0$. Then:*

(i) *R is semisimple.*

(ii) *If $x = x' \neq 0$ then $\mathbf{Q}[x]$ is a direct sum of totally real algebraic number fields.*

(iii) *If $x = x' \neq 0$, then $\sigma(yxy) \geqq 0$ for all $y \in \mathbf{Q}[x]$ if and only if x is a sum of squares in $\mathbf{Q}[x]$.*

Proof of (i): Let us show first that if $x = x' \neq 0$ then x cannot be nilpotent. If it were, we could write $x^{2^m} = 0$ but $x^{2^{m-1}} \neq 0$, and hence $x^{2^m} = (x^{2^{m-1}})^2 = 0$. Since $x^{2^{m-1}}$ is symmetric, this contradicts the strict positivity of σ. Suppose now that $x \neq 0$ is an element of a nilpotent right ideal. Then xx' is symmetric, and is in the ideal. It is therefore nilpotent, contradicting the positivity of σ.

Proof of (ii): Consider $\mathbf{Q}[x]$ with $x = x' \neq 0$. It is a commutative subalgebra of R, and the involution and trace of R induce an involution and trace in this subalgebra with similar properties. Since it is semisimple, it is a direct sum of fields. On each field F we again have an involution and trace. We have therefore only to consider the case where $R = F$. Every $\mathbf{Q}$-linear functional of F is of type $\sigma(y) = \mathrm{Tr}(\alpha y)$, where Tr is the ordinary trace from F to $\mathbf{Q}$, and α is a suitable element of F. By hypothesis we have

$$\mathrm{Tr}(\alpha y^2) = \sigma(y^2) > 0$$

if $y \neq 0$. If some conjugate of α were not real, then the corresponding conjugate of F would be dense in the complex numbers, and we could choose y in F such that αy^2 is very large negative at this conjugate, and very close to 0 everywhere else. (This is an elementary approximation theorem for valuations. For a proof, see Artin-Whaples, "Characterization of fields by the product formula," Bull. Amer. Math. Soc. 51 (1945) pp. 469–492.) We therefore get a contradiction. We see simultaneously that α is totally positive, and similarly that F is totally real.

Proof of (iii): We may again assume that $R = F$ is totally real, and that each element of F is invariant under the involution. As we have just seen, we have $\sigma(xy^2) > 0$ for every $x \neq 0$ in F, and $y \in F$, and we have also seen that in this case, x is totally positive. It is therefore a sum of squares. The converse is equally clear, and we have proved our theorem.

COROLLARY 1. *Let R be as in the theorem, and $x \in R$ be such that $x = x'$ and $\sigma(\lambda x \lambda) \geqq 0$ for all $\lambda \in \mathbf{Q}[x]$. Then for all $\beta \in R$ such that $\beta' x \beta \neq 0$, we have*

$$\sigma(\beta' x \beta) > 0.$$

Proof: We can write $x = \sum y_i^2$ with $y_i \in \mathbf{Q}[x]$. If we put $z = y_i \beta$, then one of the elements $z_i' z_i$ is not equal to 0, and

$$\sigma(\beta' x \beta) = \sum \sigma(z_i' z_i) > 0.$$

COROLLARY 2. *Let R be as in the theorem, and $\alpha = \alpha'$ be a symmetric element of R. Assume that α is a sum of squares in $\mathbf{Q}[\alpha]$. If $\lambda \in R$, and if $\lambda' \alpha \neq 0$, then $\lambda' \alpha \lambda \neq 0$.*

Proof: If we write $\alpha = \sum \beta_i^2$ with $\beta_i \in \mathbf{Q}[\alpha]$, then we must have $\lambda' \beta_i \neq 0$ for some i. Hence $0 \neq (\lambda' \beta_i)(\lambda' \beta_i)' = \lambda' \beta_i^2 \lambda$. Since $\sigma(\lambda' \beta_i^2 \lambda) \geqq 0$ for every i, and since for some i it is > 0, this proves that $\lambda' \alpha \lambda$ cannot be equal to 0.

In order to apply Theorem 3 to the algebra of endomorphisms of an abelian variety, we must now make some hypotheses on this algebra, which will be proved in the chapter on l-adic representations.

Hypothesis 1. The module $\mathrm{End}(A)$ over $\mathbf{Z}$ is of finite type, and hence every element of $\mathrm{End}_{\mathbf{Q}}(A)$ satisfies an algebraic equation over $\mathbf{Q}$.

Hypothesis 2. The trace is commutative, i.e., $\mathrm{tr}(\alpha\beta) = \mathrm{tr}(\beta\alpha)$.

Hypothesis 3. The characteristic roots of an endomorphism coincide with the characteristic roots of the characteristic polynomial of a representation of the algebra $\mathrm{End}_{\mathbf{Q}}(A)$.

THEOREM 4. *Under Hypotheses* 1 *and* 2 *made on* $\mathrm{End}_{\mathbf{Q}}(A)$, *let* $\alpha \in \mathrm{End}_{\mathbf{Q}}(A)$ *and assume* $\alpha = \alpha' \neq 0$. *Then the following conditions are equivalent*:

(i) α *is positive*;

(ii) α *is a sum of squares in* $\mathbf{Q}[\alpha]$.

Proof: If α is positive, and $\beta \in \mathbf{Q}[\alpha]$ then by Proposition 14 we have $\mathrm{tr}(\beta\alpha\beta) \geqq 0$. Hence α is a sum of squares by Theorem 3. Conversely, if $\alpha = \sum \beta_i^2 = \sum \beta_i'\beta_i$ with $\beta_i \in \mathbf{Q}[\alpha]$ then there exists an integer $e > 0$ and positive divisors Y_i such that $(e\beta_i)'(e\beta_i) = \varphi_\xi^{-1}\varphi_{Y_i}$ (Proposition 13). Let us put $Y = \sum Y_i$. Then $e^2\alpha = \varphi_\xi^{-1}\varphi_Y$ and α is positive.

COROLLARY 1. *If* α, β *are two symmetric elements of* $\mathrm{End}_{\mathbf{Q}}(A)$ *such that* $\alpha > 0$ *and* $\beta > 0$, *and* $\alpha\beta \neq 0$, *then* $\mathrm{tr}(\alpha\beta) > 0$.

Proof: Write $\alpha = \sum \lambda_i^2$ with $\lambda_i \in \mathbf{Q}[\alpha]$. We have $\mathrm{tr}(\alpha\beta) = \sum \mathrm{tr}(\lambda_i^2\beta) = \sum \mathrm{tr}(\lambda_i\beta\lambda_i) \geqq 0$. Since there exists some i such that $\lambda_i\beta \neq 0$, we find by Proposition 14 and Corollary 2 of Theorem 3 that $\lambda_i\beta\lambda_i \neq 0$ and that the strict inequality holds.

COROLLARY 2. *Let* $Y > 0$ *be a positive divisor on* A, *and* $\hat{Z} > 0$ *a positive divisor on* $\hat{A}$. *If* $\alpha = \kappa_A^{-1}\varphi_{\hat{Z}}\varphi_Y$ *and* $\alpha \neq 0$, *then* $\mathrm{tr}(\alpha) > 0$.

Proof: Let $\lambda = \varphi_X^{-1}\varphi_W$ with $W = \varphi_X^{-1}(\hat{Z})$ for some polar divisor X representing ξ, and let $\beta = \varphi_X^{-1}\varphi_Y$. The commutativity relations of Propositions 10 and 11 of § 2 show that $\alpha = \lambda\beta$, and we can then apply Corollary 1.

In view of Hypothesis 3, we see that if α is a symmetric element of $\mathrm{End}_{\mathbf{Q}}(A)$, and if ω_j are the characteristic roots of α, then the ω_j are totally real algebraic numbers, in view of the fact that $\mathbf{Q}[\alpha]$ is a direct sum of totally real fields. As an application, we can deduce properties of the norm. For every element $\alpha \in \mathrm{End}_{\mathbf{Q}}(A)$, we have:

N1. $\|\alpha\| = \|\alpha'\|$,

N2. $\|\alpha'\alpha\| = \|\alpha\alpha'\|$,

N3. If $\alpha \neq 0$, and $\beta \neq 0$ then $\|\alpha\beta\| < \|\alpha\| \, \|\beta\|$.

The first two conditions come from the commutativity of the trace. As to the third, we have

$$\|\alpha\beta\|^2 = \operatorname{tr}(\beta'\alpha'\alpha\beta) = \operatorname{tr}(\alpha'\alpha\beta\beta') = (\alpha'\alpha,\ \beta\beta') \leqq \|\alpha'\alpha\|\ \|\beta\beta'\|,$$

this last inequality being the Schwartz inequality. We are therefore reduced to proving the inequality $\|\alpha'\alpha\| < \|\alpha\|^2$. By definition, we have $\|\alpha'\alpha\|^2 = \operatorname{tr}(\alpha'\alpha\alpha'\alpha) = \operatorname{tr}((\alpha'\alpha)^2)$. Let ω_j be the characteristic roots of $\alpha'\alpha$. Then ω_j^2 are the characteristic roots of $(\alpha'\alpha)^2$, according to Theorem 8 of Chapter IV, § 3. We note that $\alpha'\alpha$ is symmetric and positive, and hence that the ω_j are totally real, totally positive algebraic numbers. Our inequality is equivalent with the assertion $\sum \omega_j^2 < (\sum \omega_j)^2$ which is true, in view of this fact.

Historical note:

The transpose of the exact sequence is due to Chow [15] who proves in fact a more general result, for arbitrary varieties.

The commutative diagrams of § 2 are due to Weil, who gives here a very elegant formalism. A large number of these results were already hidden in his treatise [85], under the disguise of l-adic matrices. This was due to the fact that the Picard variety was not available.

Lang [48] has applied the formalism of § 2 and shows how one recovers in this manner the theorem $\sigma(\alpha'\alpha) > 0$ of Weil, who was the first to recognize that Castelnuovo's theorem on the equivalence defect of correspondences on a curve could be expressed as a theorem on abelian varieties. As mentioned already, we shall recover Castelnuovo's theorem in due course. For further remarks of a general nature concerning the zeta function, see our discussion preceding the bibliography. The formalism of the divisor class $D_\xi(\alpha,\ \beta)$ has been copied from [48]. Note that Proposition 2 of Chapter IV, § 1 which lies at the beginning of this formalism gives us on abelian varieties the analogue of Hasse's norm formula [35] for elliptic curves. Note also that on a Jacobian, the involution $\alpha \to \alpha' = \varphi_\Theta^{-1}\ {}^t\alpha\varphi_\Theta$ is none other than the Rosati anti-automorphism.

The theorems of § 3 are applied in the theory of complex

multiplication [78], [81], [96]. Some of the consequences of the positive definiteness of the trace (Theorems 3, 4 and their corollaries) systematize and generalize results of Weil [85] and Morikawa [63]. The exposition is based on [48], and we have replaced a comparatively deep fact of number theory used by Morikawa by a trick of Albert's [1].

We note finally that Morikawa's paper [63] includes an algebraic proof of the Riemann-Roch theorem on abelian varieties, in characteristic 0. His proof fails in characteristic p for lack of information concerning infinitesimal points. Special techniques in characteristic p are presumably needed to deal with this case. (The Riemann-Roch theorem on an abelian variety states that if X is a positive non-degenerate divisor on A, then $l(X) = \sqrt{\nu(\varphi_X)}$.)

CHAPTER VI

The Picard Variety of an Arbitrary Variety

We are going to see in this chapter how one can prove the existence of the Picard variety of a variety V, complete and non-singular in codimension 1 in a simple manner, once one knows the existence of the Picard variety of an abelian variety. The Picard variety of V is derived functorially from that of its Albanese variety, and we shall use this fact to get the theory of divisorial correspondences on a product $U \times V$. As a special case, we obtain the theory of correspondences on a curve, which gives us the Lefschetz fixed point formula. The group of correspondence classes of the curve is isomorphic to the group of endomorphisms of its Jacobian. The characteristic polynomial of an endomorphism corresponds to the characteristic polynomial of the representation of the endomorphism on the first homology group (in the classical case), and its trace is the trace of this representation. Combining the Lefschetz fixed point formula with the results of Chapter V, we obtain in a natural way the Riemann hypothesis for curves.

§ 1. *Construction of the Picard variety*

Let V be a variety, and $f : V \to A$ a rational map into an abelian variety. Let $(\hat{A}, D)$ be a Picard variety of A.

$$V \xrightarrow{f} A \overset{D}{\times} \hat{A}.$$

It may be that the composed divisor $D \circ f$ is not defined. However, we know from the Appendix, § 2 that we can always find a divisor D_1 linearly equivalent to D such that $D_1 \circ f$ is defined. More generally, if D_1, D_2 are in the same correspondence class as D on $A \times \hat{A}$, and if $D_1 \circ f$ and $D_2 \circ f$ are defined, then

$$f^{-1}[{}^tD_1(w) - {}^tD_1(z)] \sim f^{-1}[{}^tD_2(w) - {}^tD_2(z)]$$

for any two generic points w, z of $\hat{A}$ (not necessarily independent).

In order to get the composed divisor, we could also change D by a translation (a, b) on $A \times \hat{A}$. Since $D_{(a,b)}$ is algebraically equivalent to D on $A \times \hat{A}$, the product theorem shows that this translation is in the same correspondence class as D.

We could also make a translation on f. More precisely, we have the following criterion.

PROPOSITION 1. *Let $f: V \to A$ be a rational map of a variety into an abelian variety, and let f_a be the translate of f (i.e., the map such that $f_a(P) = f(P) + a$). Then if $X \in D_a(A)$ and if $f^{-1}(X)$ and $f_a^{-1}(X)$ are defined, they are linearly equivalent.*

Proof: The biholomorphic transformation $(P, u) \to (P, u + a)$ of $V \times A$ onto itself shows that $f_a^{-1}(X) = f^{-1}(X_{-a})$. Since X is assumed algebraically equivalent to 0, we know that $X_{-a} \sim X$ by Proposition 4 of Chapter III, § 3, and our assertion is a consequence of Corollary 3 of Theorem 3, Appendix, § 1.

After these remarks, the next theorem shows us how to obtain the Picard variety of a variety from that of its Albanese variety.

THEOREM 1. *Let V be a variety, complete and non-singular in codimension 1, defined over a field k. Let $f: V \to A$ be a canonical map of V into its Albanese variety, also defined over k. Let $(\hat{A}, D)$ be a Picard variety of A defined over k, and such that the composed divisor $E = D \circ f$ is defined. Then $(\hat{A}, E)$ is a Picard variety of V, defined over k.*

Proof: We are first going to show that we may assume that V has a simple rational point over k. Indeed, condition (1) in the definition of the Picard variety is obviously satisfied over k, because it is independent of a field of definition. As to condition (2), suppose that $X \in D_a(V)$ is rational over a field K. Let P be a generic point of V over K, and put $k' = k(P)$. Assume that the theorem is true relative to k'. Then if we write

$$X \sim {}^tE(w + z) - {}^tE(z)$$

with w, $z \in \hat{A}$ and z generic over Kk', we have w rational over $K(P)$. Since we can take two independent generic points P, Q

of V over K, it follows that w is rational over $K(P) \cap K(Q) = K$.

We assume therefore that V has a simple rational point over k. We go back to the last section of the chapter on the theorem of the square, and apply Theorem 5 of Chapter III, § 4 to the divisor $E = D \circ f$ on $V \times \hat{A}$. We must show that taking the quotient as in (iv) of that theorem, we retain $\hat{A}$, i.e., that the map φ is birational.

From Theorem 7 of the Appendix, § 2 we get

$$E(P) - E(Q) \sim D[f(P)] - D[f(Q)] \tag{1}$$

for P, Q generic independent on V. Taking $G = \hat{A}$ in (iv) of Theorem 5, Chapter III, § 4, and letting $\varphi : \hat{A} \to B$ be as in this theorem, we apply again the theory of correspondences of the appendix to the sequence

$$\hat{A} \xrightarrow{\varphi} B \overset{Y}{\times} V$$

where Y is the quotient divisor on $B \times V$, i.e., is such that for P, Q generic independent on V,

$$\varphi^{-1}[{}^tY(P) - {}^tY(Q)] \sim E(P) - E(Q). \tag{2}$$

Denote by $g : V \times V \to \hat{B}$ the admissible rational map defined by

$$g(P, Q) = \mathrm{Cl}\,[{}^tY(P) - {}^tY(Q)].$$

From (1) and (2) above, and the definition of the Poincaré divisor, we get

$${}^t\varphi g = \kappa_A F$$

where $F : V \times V \to A$ is the rational map defined by $F(P, Q) = f(P) - f(Q)$. We note that F is admissible.

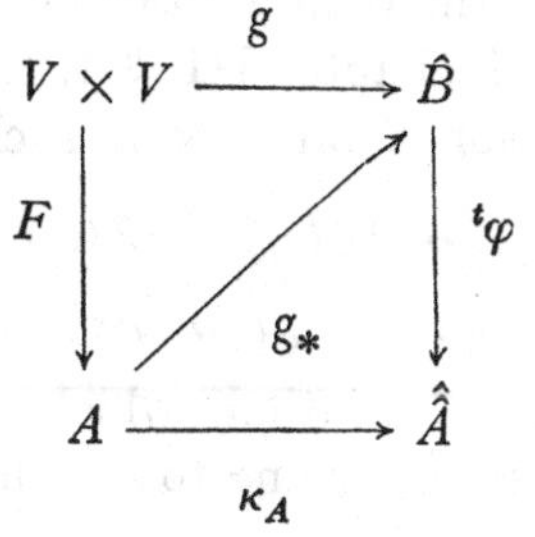

Let $g_* : A \to \hat{B}$ be the homomorphism of the Albanese variety induced by g. Then

$${}^t\varphi g_* = \kappa_A.$$

Take the transpose of this equation. Then ${}^t\kappa_A = {}^t g_* {}^t({}^t\varphi)$. From Proposition 9 of Chapter V, § 2 we have $\nu({}^t\kappa_A) = 1$ and hence $\nu[{}^t({}^t\varphi)] = 1$. But we also know that $\nu(\kappa_{\hat{A}}) = 1$, and we have the commutative diagram

$$\begin{array}{ccc} \hat{A} & \xrightarrow{\ \varphi\ } & B \\ {\scriptstyle \kappa_{\hat{A}}}\downarrow & & \downarrow{\scriptstyle \kappa_B} \\ \hat{\hat{A}} & \xrightarrow[\ t^2\varphi\ ]{} & \hat{\hat{B}} \end{array}$$

We see therefore that $\nu(\varphi) = 1$, or in other words that φ is a birational isomorphism. We could have taken $B = \hat{A}$ and $Y = D \circ f$. If we do this, we see already that $(\hat{A}, D \circ f)$ satisfies the first three conditions of Theorem 9, Chapter IV, § 4.

All that remains to be done is to prove the last condition, or equivalently, to prove that the mapping

$$w \to \mathrm{Cl}\,[{}^tE(w + z) - {}^tE(z)]$$

is surjective. We rely on the following lemma.

LEMMA 1. *Let V be an arbitrary variety, X a divisor in $D_a(V)$. Then there exists an abelian variety B, a rational map $h : V \to B$, and a divisor $Y \in D_a(B)$ such that $X \sim h^{-1}(Y)$.*

Proof: According to the definition of algebraic equivalence, or rather Proposition 1 of Chapter III, § 1, there exists an abelian variety B and a divisor Z on $V \times B$ such that

$$X \sim {}^tZ(v_1) - {}^tZ(v_2)$$

with two generic points v_1, v_2 of B. (We can make them generic by making a generic translation, which does not change the linear equivalence class, according to the theorem of the square.)

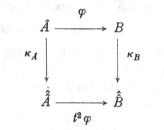

For P, Q generic independent on V, the mapping

$$(P, Q) \to \mathrm{Cl}\,[Z(P) - Z(Q)]$$

gives an admissible rational map of $V \times V$ into $\hat{B}$. Let $g : V \to \hat{B}$ be a map such that $g(P) - g(Q) = \mathrm{Cl}\,[Z(P) - Z(Q)]$, and let E be a Poincaré divisor on $B \times \hat{B}$:

$$V \xrightarrow{g} \hat{B} \overset{{}^tE}{\times} B.$$

By definition, we may write

$$Z(P) - Z(Q) \sim {}^tE(w + z) - {}^tE(z)$$

with z generic on $\hat{B}$, and we have $w = g(P) - g(Q)$. This shows that ${}^tE \circ g$ is in the same correspondence class on $V \times B$ as Z, and hence by Theorem 7 of the Appendix, § 2 we get

$${}^tZ(v_1) - {}^tZ(v_2) \sim g^{-1}[E(v_1) - E(v_2)].$$

We take for Y the divisor $E(v_1) - E(v_2)$ on $\hat{B}$, and we take g for h. This proves our lemma.

Let $f : V \to A$ be a canonical map of V into its Albanese variety, V being now assumed again complete and non-singular in codimension 1. If $h_* : A \to B$ is the induced homomorphism (with h as in Lemma 2) we have $h = h_* f + b$ with a constant $b \in B$.

$$V \xrightarrow{f} A \xrightarrow{h_*} B.$$

Using Theorem 2 of Appendix, § 1 and Proposition 1, we obtain, after changing Y by a linear equivalence if necessary, or by a generic translation,

$$h^{-1}(Y) \sim f^{-1}(h_*^{-1}(Y)).$$

This shows that we can find a divisor $Y_1 \in D_a(A)$ such that $X \sim f^{-1}(Y_1)$. Using the definition of the composed divisor $E = D \circ f$ on $V \times \hat{A}$, we see that there exist two points w, z of $\hat{A}$ with z generic over $k(w)$ such that

$$X \sim {}^tE(w + z) - {}^tE(z).$$

This proves the surjectivity, and concludes the proof of Theorem 1.

From now on, we can identify Pic (V) with $\hat{A}$ by means of the isomorphism

$$w \to \mathrm{Cl}\,[{}^tE(w+z) - {}^tE(z)].$$

Let U, V be two varieties, complete and non-singular in codimension 1, and let $f : U \to A$ and $g : V \to B$ be two canonical maps into their Albanese varieties. Let $h : U \to V$ be a rational map of U into V such that $h(U)$ is simple on V. Then there is an induced homomorphism $h_* : A \to B$. If V has singularities, then in general the inverse image $h^{-1}(Y)$ of a divisor $Y \in D_a(V)$ will not necessarily be in $D_a(U)$. But if V is non-singular, then we have a homomorphism

$$h^* : \mathrm{Pic}\,(V) \to \mathrm{Pic}\,(U)$$

defined by $Y \to h^{-1}(Y)$, using Proposition 1 of Chapter V, § 1. Using Theorem 2 of the Appendix, § 1 we see that the transpose ${}^th_* : \hat{B} \to \hat{A}$ gives precisely the same homomorphism on the divisor classes, provided that the diagram

$$\begin{array}{ccc} U & \xrightarrow{\;h\;} & V \\ {\scriptstyle f}\downarrow & & \downarrow{\scriptstyle g} \\ A & \xrightarrow[\;h_*\;]{} & B \end{array}$$

is commutative. The only thing which can prevent it from being commutative is the fact that we may have to add a constant to h_*. But in this case, we can apply Proposition 1.

Finally, a word about the birational invariance of the Picard variety. Since the Albanese variety is obviously a birational invariant, we interpret the fact that the Picard group of a variety is obtained by taking the inverse image of the Picard group of its Albanese variety as a statement of the birational invariance. If both U, V are non-singular, then, of course, a birational correspondence between them induces directly the isomorphisms between their Picard groups.

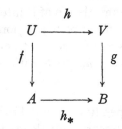

§ 2. *Divisorial correspondences*

Let U, V be two varieties, complete and non-singular in codimension 1. For the rest of this section, we let

$$\varphi : U \times U \to A \quad \text{and} \quad \psi : V \times V \to B$$

be two canonical admissible maps of $U \times U$ and $V \times V$, respectively, into their Albanese varieties. We assume that U, V, φ, ψ, A, B are defined over a field k, and that $(\hat{A}, D)$ and $(\hat{B}, E)$ are Picard varieties of A and B defined over k. Let

$$f : U \to A \quad \text{and} \quad g : V \to B$$

be canonical maps of U, V, respectively, into their Albanese varieties and assume that they are also defined over k.

Let X be a divisor on $U \times V$ rational over k. Then X induces a rational map

$$h_X : U \times U \to \hat{B}$$

by the expression

$$h_X(P, Q) = \text{Cl}\,[X(P) - X(Q)]$$

for P, Q generic independent. It is obvious that h_X is admissible. Hence there is an induced homomorphism $h_{X*} : A \to \hat{B}$ which will be denoted by λ_X.

Conversely, suppose that we are given a homomorphism $\lambda : A \to \hat{B}$ defined over k. We can then define a rational map $h : U \times U \to \hat{B}$ by the equation $h = \lambda\varphi$. There exists a divisor T on $A \times B$ such that $\lambda = \lambda_T$ in the sense of Chapter V, § 2. That is to say, for u, v generic independent on A, we have

$$\lambda u = \text{Cl}\,[T(u + v) - T(v)].$$

Let $F : U \times V \to A \times B$ be the product mapping (f, g) and put $X = F^{-1}(T)$. We can assume that this expression is defined, after changing T by a generic translation. Let P be a generic point V over k. The following diagram is commutative:

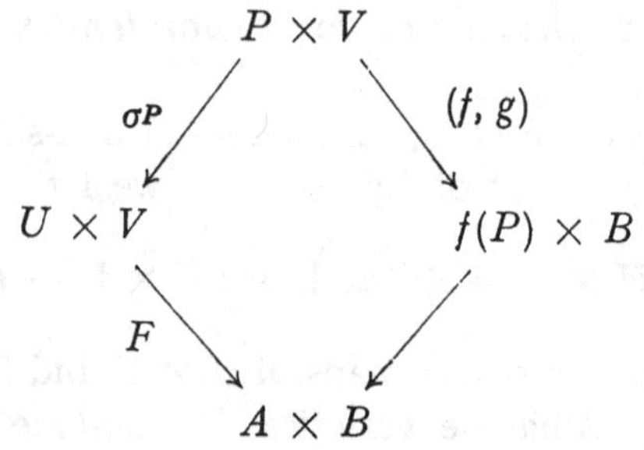

with σ_P equal to the inclusion. Formally, we see that

$$F^{-1}(T) \cdot (P \times V) = P \times g^{-1}[T(f(P))].$$

If $U \times V$ is non-singular, we get what we want by Theorem 2 of the Appendix, § 1. If not, let us justify our steps. For the expression on the left, we have

$$\mathrm{pr}_{U\times V}\,[\Gamma_F \cdot (U \times V \times T)] \cdot (P \times V)$$

and since $A \times B$ is non-singular, and the intersection

$$[\Gamma_F \cdot (U \times V \times T)] \cdot (P \times V \times A \times B)$$

is defined, this is equal to

$$\mathrm{pr}_{U\times V}\{[(U \times V \times T) \cdot \Gamma_F] \cdot (P \times V \times A \times B)\}.$$

Using associativity, and the fact that Γ_F is obtained from $(\Gamma_f \times V \times B) \cdot (U \times A \times \Gamma_g)$ by transforming this cycle under the permutation of the two factors in the middle, we see that our cycle is equal to

$$\mathrm{pr}_{U\times V}\{(U \times V \times T) \cdot [(P \times V \times f(P) \times B) \cdot (U \times A \times \Gamma_g)']\}$$

where the $'$ means that we have to permute the two middle factors. Since $F^{-1}(T)$ is defined, it follows that $T(f(P))$ is defined, and we can apply associativity on the left. We find

$$\mathrm{pr}_{U\times V}\{[P \times f(P) \times V \times T(f(P))] \cdot (U \times A \times \Gamma_g)\}$$

after a reorganization of the factors. We thus get the term on the right-hand side of our original expression.

We have therefore shown that

$$\mathrm{Cl}\,[X(P) - X(Q)] = \mathrm{Cl}\,[g^{-1}[T(f(P)) - T(f(Q))]].$$

In view of the identification of the Picard variety of V with that of its Albanese variety in § 1, this means that

$$h_X(P, Q) = \lambda_T \varphi(P, Q).$$

The following diagram is commutative.

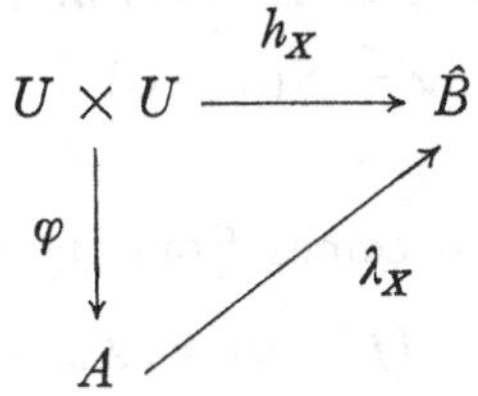

Since P, Q are generic independent, we get $\lambda_X = \lambda_T$. We may summarize our results in the following manner.

THEOREM 2. *Let U, V be complete and non-singular in codimension* 1. *Let (A, f) and (B, g) be Albanese varieties of U and V respectively, let X be a divisor on $U \times V$ and let $\lambda_X : A \to \hat{B}$ be the homomorphism induced by the rational map*

$$(P, Q) \to \mathrm{Cl}\,[X(P) - X(Q)]$$

of $U \times U$ into $\hat{B}$. Then the mapping $X \to \lambda_X$ induces an isomorphism of the correspondence classes of $U \times V$ onto $H(A, \hat{B})$. If in addition T is a divisor on $A \times B$, if $F : U \times V \to A \times B$ is the product (f, g), and if $X = F^{-1}(T)$ is defined, then $\lambda_X = \lambda_T$, this last homomorphism being defined by

$$\lambda_T u = \mathrm{Cl}\,[T(u + v) - T(v)].$$

§ 3. *Application to the theory of curves*

Let C be a complete non-singular curve, and $f : C \to J$ a canonical map of C into its Jacobian, defined over a field k. We may assume that it is the inclusion. As usual, we denote by Θ the divisor $C \oplus C \oplus \ldots \oplus C$ taken $g - 1$ times on J. We recall that $s_2 : J \times J \to J$ is the sum. The next theorem identifies J and $\hat{J}$ by the map $\varphi_\Theta : J \to \hat{J}$.

THEOREM 3. *The Jacobian of a curve is self-dual. More precisely, if we put* $D = - s_2^{-1}(\Theta)$, *then* D *is a Poincaré divisor on* $J \times J$, *and* (J, D) *is a Picard variety of* J. *If* $X \in D_a(J)$, *then there exists a point* $x \in J$ *such that*

$$X \sim \Theta_x - \Theta,$$

uniquely determined by Cl (X). *This point* x *is rational over the smallest field of rationality of* X *containing* k, *and we have*

$$x = S(X_u \cdot C)$$

for any generic point u *of* J.

Proof: The minus sign comes from the fact that

$$s_2^{-1}(\Theta) \cdot (J \times u) = \Theta_{-u} \times u.$$

The existence of the point x comes from a result obtained some time ago, namely Theorem 4 of Chapter IV, § 2. To prove the uniqueness, let u be a generic point of J over $k(x)$, and consider the intersection

$$(W_{u+x} - W_u) \cdot C$$

with $W = \Theta^-$. We know that $W \equiv \Theta$ by the corollary of Proposition 2 of Chapter IV, § 1 and consequently $W_{u+x} - W_u$ is linearly equivalent to $\Theta_{u+x} - \Theta_u$. By Proposition 3 of Chapter III, § 1, and by Abel's theorem, we conclude from Proposition 5 of Chapter II, § 2 that

$$\begin{aligned} S[(W_{u+x} - W_u) \cdot C] &= S(W_{u+x} \cdot C) - S(W_u \cdot C) \\ &= u + x - u \\ &= x. \end{aligned}$$

On the other hand, we have $X_u \sim X$ by Proposition 4 of Chapter III, § 3 and if X is rational over $K \supset k$, then X_u is rational over $K(u)$, and $X_u \sim W_{u+x} - W_u$. Hence we get

$$S(X_u \cdot C) = S[(W_{u+x} - W_u) \cdot C] = x.$$

This shows that x is rational over $K(u)$ for all generic points u of J over K, and hence is rational over K.

A similar argument also gives us the following result.

THEOREM 4. *Let* $\alpha: J \to A$ *be a homomorphism, and* X *a divisor on* A. *Let* u, v *be two independent generic points of* A. *Then the homomorphism* $\varphi_\Theta^{-1}\, {}^t\alpha\, \varphi_X$ *is given by the mapping*

$$u \to S[\alpha^{-1}(X_{u+v} - X_v) \cdot C].$$

Proof: Since $X_{u+v} - X_v \in D_a(A)$, we have $\alpha^{-1}(X_{u+v} - X_v) \sim W_{t+z} - W_z$ for some $t \in J$ and z generic. Hence

$$S[C \cdot \alpha^{-1}(X_{u+v} - X_v)] = S[C \cdot (W_{t+z} - W_z)] = t.$$

This proves our theorem, if we take the definitions into account.

We apply Theorem 3 to show that a Jacobian variety has no torsion. More precisely, we prove here the following fact.

PROPOSITION 2. *Let* J *be a Jacobian. Then there is no element of period* p *in the torsion group* $D_\tau(J)/D_a(J)$.

Proof: We shall use the fact proved in Theorem 3 that the Picard variety of C is isomorphic to that of J under the inverse mapping, which in this case is the intersection.

LEMMA 2. (Serre) *Let* U, V *be two complete non-singular varieties. Let* $f: U \to V$ *be an everywhere-defined, surjective, separable map. Let*

$$f^{-1}: D(V)/D_l(V) \to D(U)/D_l(U)$$

be the induced map on the divisor classes for linear equivalence. Then there is no element of order p *(the characteristic) in the kernel of* f^{-1}.

Proof: Let X be a divisor on V and φ a function on V such that $pX = (\varphi)$. Suppose that there is a function ω on U such that $f^{-1}(X) = (\omega)$. Then $f^{-1}(pX) = p \cdot f^{-1}(X) = (\omega^p)$. If we denote by φ_1 the function $\varphi \circ f$ on U, we see that ω^p and φ_1 have the same divisor, and hence differ by a constant. We could have chosen ω such that $\omega^p = \varphi_1$. Since the function field of U is separable over that of V by hypothesis, it follows that φ must already be a pth power in the function field of V, and hence that X is linearly equivalent to 0.

We shall now prove that there is no p-torsion. The case of torsion prime to p will be settled by other methods, for all abelian varieties, in Chapter VII. Let $f: C \to J$ be a canonical map of C into its Jacobian, which we may assume to be the inclusion. Let f_g be the product of f with itself g times, and let s_g be the sum. Let $F = s_g f_g$:

$$C \times \ldots \times C \xrightarrow{f_g} J \times \ldots \times J \xrightarrow{s_g} J.$$

We know that if $X \equiv 0$ on J, then $s_g^{-1}(X) \sim \sum p_i^{-1}(X)$ by Theorem 2 of Chapter IV, § 1. Put $\mathfrak{a} = X \cdot C$, which we may assumed to be defined (after a generic translation on X if necessary). If $pX \sim 0$ on J, then deg $(\mathfrak{a}) = 0$, and $\mathfrak{a}$ is algebraically equivalent to 0 on C. Since Pic (J) and Pic (C) are isomorphic under the intersection mapping, there exists $X_1 \in D_a(J)$ such that $X_1 \cdot C \sim \mathfrak{a}$, and such that Cl (X_1) has period p in $\hat{J}$. Put $Y = X - X_1$. Then $Y \cdot C = \mathfrak{b}$ is linearly equivalent to 0 on C and $pY \sim 0$ on J. It will suffice to prove that $Y \sim 0$. We have

$$F^{-1}(Y) \sim \sum (C \times \ldots \times \mathfrak{b} \times \ldots \times C)$$

which is linearly equivalent to 0 on the product of C with itself g times. According to Lemma 2 we must have $Y \sim 0$. This proves our proposition.

We can define an isomorphism of Pic (C) into Pic (J) which is inverse to the one of Theorem 3. Let $\mathfrak{a} = \sum m_j P_j$ be a divisor of degree 0 on C, and denote by $\Theta(\mathfrak{a})$ the divisor $\sum m_j \Theta_{f(P_j)}$. Then the mappings

$$\mathfrak{a} \to \Theta(\mathfrak{a})$$

and

$$X \to X \cdot C$$

for $X \in D_a(J)$ such that the intersection is defined, induce isomorphisms on the Picard groups, which are inverse to each other. In particular, if $\mathfrak{a} \sim 0$ on C, we can give explicitly the function on J belonging to $\Theta(\mathfrak{a})$.

THEOREM 5. *Let f: $C \to J$ be a canonical map of a complete non-singular curve into its Jacobian, and let*

$$\mathfrak{a} = \sum m_j P_j = (\varphi)$$

be a divisor on C, linearly equivalent to 0. Let k be a field of defini-

tion for f, over which $\mathfrak{a}$ is rational, and φ is defined. Let $\mathfrak{m} = \sum_{i=1}^{g}(M_i)$ be a divisor on C such that the points M_i are generic independent over k, and put $u = \sum f(M_i)$. Then there exists a function ω on J defined over k, such that

$$\omega(u) = \prod \varphi(M_i)$$

and we have

$$(\omega) = \sum m_j \Theta_{f(P_j)}.$$

Proof: It is obvious that the product of the $\varphi(M_i)$ is invariant under the permutations of the M_i, and hence that the function ω exists. Let C^g be the product of C with itself g times. The divisor on C^g of the function $\Phi(M_1, \ldots, M_g) = \prod\varphi(M_i)$ is equal to $\sum (C \times \ldots \times \mathfrak{a} \times \ldots \times C)$. According to the norm theorem (F—VIII_2 Th. 7) the divisor of the norm of Φ, from C^g to J, is obtained by taking the image of the divisor (Φ) by the function $F : C^g \to J$ defined as before. Since Φ is already invariant by the automorphisms of C^g over J, its norm is equal to $\omega^{g!}$ and we get

$$g!(\omega) = (\omega^{g!}) = g! \sum m_j \Theta_{f(P_j)}.$$

This proves our theorem. The reader will note the analogy between our preceding arguments and those of Lemma 2, Chapter IV, § 2.

Let us develop the theory of correspondences on the curve C. Let X, Y be two divisors on $C \times C$. They are cycles of dimension 1 on a surface. If $X \cdot Y$ is defined on $C \times C$, it is therefore of dimension 0. If $X \cdot Y$ is not defined, and if X is a variety, we can find a function φ on $C \times C$ such that if we put $Y_1 = Y + (\varphi)$ then $X \cdot Y_1$ is defined (Appendix, § 1, Proposition 3), and in this case the degree $\deg (X \cdot Y_1)$ is independent of the auxiliary function selected. In order to prove this last statement, we must show that if $Y \sim 0$ on $C \times C$ then $\deg (X \cdot Y) = 0$ for every subvariety X of dimension 1 of $C \times C$. But φ induces on X a function which is not the constant 0, according to F—VIII_2 Th. 3. By F—VIII_2 Th. 7 it follows that $\text{pr}_1 [X \cdot (\varphi)]$ is linearly equivalent to 0 on C, and hence of degree 0. Since the projection preserves the degree of a cycle of dimension 0, we have proved our assertion.

We can now define numerical equivalence on $C \times C$ just as we did for an abelian variety in Chapter IV. The preceding remark gives us the technical equivalent of a generic translation on an abelian variety. If $X = \sum n_i X_i$ is an expression of X as a formal sum of subvarieties of $C \times C$, we can define the symbol

$$I(X \cdot Y) = \sum n_i I(X_i \cdot Y)$$

by linearity, if we take for $I(X_i \cdot Y)$ the degree of the cycle $X_i \cdot Y_1$ for any divisor Y_1 on $C \times C$ linearly equivalent to Y such that $X_i \cdot Y_1$ is defined. One shows immediately that

$$I(X + X', Y) = I(X, Y) + I(X', Y),$$
$$I(X \cdot Y) = I(Y \cdot X).$$

We say that X is numerically equivalent to 0 if $X \cdot Y$ has degree 0 for any Y such that this intersection is defined. The intersection number $I(X \cdot Y)$ gives a bilinear form on the numerical equivalence classes of divisors on $C \times C$.

Let P be a point of C such that $X \cdot (P \times C)$ is defined. The degree of this cycle does not depend on the point P selected, by the principle of conservation of number F—VII$_6$ Th. 13. We denote it by $d(X)$, and we shall denote by $d'(X)$ the integer $d({}^tX)$, equal to the degree of $X \cdot (C \times P)$ for every generic point P. We contend that the function

$$\sigma(X) = d(X) + d'(X) - I(X \cdot \Delta)$$

(Δ being the diagonal) is a function of the correspondence classes on $C \times C$, i.e., that if X is a trivial correspondence of type $(\varphi) + \mathfrak{a} \times C + C \times \mathfrak{b}$, then $\sigma(X) = 0$. Indeed, let P be a generic point of C. Then $(\mathfrak{a} \times C) \cdot (P \times C) = 0$ and $(C \times \mathfrak{b}) \cdot (P \times C) = P \times \mathfrak{b}$. The definitions show immediately that

$$d(X) = \deg(\mathfrak{b}) \quad \text{and} \quad d'(X) = \deg(\mathfrak{a}),$$

always taking into account the fact that the divisor of a function on a curve has degree 0. Since

$$\deg \{[(\mathfrak{a} \times C) + (C \times \mathfrak{b})] \cdot \Delta\} = \deg(\mathfrak{a}) + \deg(\mathfrak{b}),$$

we find $\sigma(X) = 0$.

We leave to the reader the verification of the fact that the functions $d(X)$, $d'(X)$, and $I(X \cdot \Delta)$ are linear in X. From this we conclude that $\sigma(X)$ is a linear functional on the correspondence classes on $C \times C$, with values in the integers $\mathbf{Z}$.

The number $I(X \cdot \Delta)$ is interpreted as the number of fixed points of the correspondence, and we have

$$I(X \cdot \Delta) = d(X) - \sigma(X) + d'(X).$$

The Lefschetz fixed point formula consists in interpreting $\sigma(X)$ as the trace of the representation of the correspondence in the first homology group (in the classical case). In our algebraic theory, we can express the Lefschetz formula as follows.

THEOREM 6. *Let C be a complete non-singular curve, X a divisor on $C \times C$, and $\tau : J \to J$ the endomorphism induced by X on the Jacobian, identified with its dual $\hat{J}$. Let ζ be the numerical equivalence class of C on J, and θ the numerical equivalence class of Θ. Then*

$$\sigma(X) = \deg\, [\zeta \cdot D_\theta(\tau)] = \mathrm{tr}(\tau)$$

where, as usual, we have $D_\theta(\tau) = (\tau + \delta)^{-1}(\theta) - \tau^{-1}(\theta) - \theta$.

Proof: Let us consider the following diagram of mappings:

$$\begin{array}{ccccccc}
C \times C & \xrightarrow{F} & J \times J & \xrightarrow{(\tau,\ \delta)} & J \times J & \xrightarrow{s_2} & J \\
\uparrow {\scriptstyle h_i} & & \uparrow {\scriptstyle H_i} & & & & \\
C & \xrightarrow[f]{} & J & & & &
\end{array}$$

Here, $f : C \to J$ is a canonical map, and we assume that there is a point P_0 of C such that $f(P_0) = 0$.

$F : C \times C \to J \times J$ is the product of f with itself.

h_i $(i = 1, 2, 3)$ is the mapping such that

$$h_1(P) = P_0 \times P, \qquad h_2(P) = P \times P_0,$$
$$h_3(P) = (P,\ P).$$

H_i $(i = 1, 2, 3)$ is the homomorphism such that

$$H_1(x) = (0, x), \qquad H_2(x) = (x, 0),$$
$$H_3(x) = (x, x).$$

The square is then commutative.

(τ, δ) is the homomorphism product of τ and the identity. s_2 is the sum.

Let u be a generic point of J, and take the inverse images of $-\Theta_u$. We may use Theorem 2 of the Appendix, § 1 and Proposition 7 of Chapter I, § 2 to guarantee that the inverse of the composed mappings is the composed mapping of the inverses.

By Theorem 3, we know that $s_2^{-1}(-\Theta_u)$ is a Poincaré divisor on $J \times J$.

By Proposition 6 of Chapter V, § 2 the divisor

$$T = (\tau, \delta)^{-1} s_2^{-1}(-\Theta_u)$$

is a correspondence on $J \times J$ such that $\lambda_T = \tau$.

According to Theorem 2 of § 2, the divisor $F^{-1}(T) = Y$ is a correspondence on $C \times C$ whose induced endomorphism on the Jacobian is equal to τ. Hence $F^{-1}(T)$ is in the same correspondence class as X.

Let us compute the degrees of the inverse images on C. As usual, we may assume that f and F are inclusion mappings.

On the one hand, we have

$$h_1^{-1}(Y) = Y \cdot (P_0 \times C).$$

On the other hand, if we go around the square the other way, and if we observe that $s_2(\tau, \delta)H_1 = \delta$, we get

$$f^{-1} H_1^{-1}(\tau, \delta)^{-1} s_2^{-1}(-\Theta_u) = (-\Theta_u) \cdot C.$$

Consequently, we have shown that

$$d(Y) = -\deg(\Theta_u \cdot C).$$

Similarly, one proves that

$$d'(Y) = -\deg(\tau^{-1}(\Theta_u) \cdot C),$$
$$I(Y \cdot \Delta) = -\deg[(\tau + \delta)^{-1}(\Theta_u) \cdot C].$$

This gives us $\sigma(Y)$. Since Y and X are in the same correspondence class,

$$\sigma(X) = \deg\{[(\tau + \delta)^{-1}(\Theta_u) - \tau^{-1}(\Theta_u) - \Theta_u] \cdot C\}.$$

We have proved the formula of the theorem, taking into account the corollary of Theorem 7, Chapter IV, § 3.

If we combine the Lefschetz fixed point formula with the Riemann hypothesis for abelian varieties, we recover the Riemann hypothesis for curves. Indeed, let C be defined over the finite field k with q elements. Then there exists a canonical admissible map $\varphi : C \times C \to J$ defined over k because k is perfect (Theorem 12 of Chapter II, § 3). The Frobenius correspondence which to each point P of C assigns the point $P^{(q)}$ obviously induces the Frobenius endomorphism $\pi : u \to u^{(q)}$ on the Jacobian. If X denotes the Frobenius correspondence on $C \times C$, then

$$d(X) = 1 \quad \text{and} \quad d'(X) = q,$$

by Lemma 1 of Chapter V, § 3 taking $r = 1$. The statement of Theorem 6 is geometric: it does not depend on the ground field over which we work. We may therefore extend our ground field in such a way that we get a canonical map $f : C \to J$. We then find

$$I(X \cdot \Delta) = 1 + q - \operatorname{tr}(\pi).$$

We have seen in Chapter V, § 3 that

$$\operatorname{tr}(\pi) = \sum_{i=1}^{2g} \omega_i$$

if g is the genus of the curve, and that $|\omega_i| = q^{1/2}$. On the other hand, one verifies without difficulty that X and Δ are transversal on $C \times C$. (Heuristically, the derivative of X is everywhere 0, so that X is everywhere horizontal, and is thus obviously transversal to the diagonal.) Consequently, the points of the intersection $X \cdot \Delta$ are the points of $X \cap \Delta$ taken with multiplicity 1, i.e., the points such that $P^{(q)} = P$. They are the rational points of the curve over k. The Lefschetz formula gives us the number of rational points on the curve. If we replace π by π^n, we find the number of rational points over the field k_n, unique extension of k of degree n. For $n \to \infty$ the number of these points is therefore equal to

$$q^n + O(q^{n/2}).$$

In particular, as soon as n is sufficiently large, there exists a rational point of C in k_n.

We also know enough to derive trivially the zeta function of C. If N_n is the number of rational points of C in k_n then

$$N_n = 1 + q^n - \sum \omega_i^n.$$

From the logarithmic derivative

$$\frac{d}{dt} \log Z(t) = \sum_{n=1}^{\infty} N_n t^{n-1}$$

we see that

$$Z(t) = \frac{F(t)}{(1-t)(1-qt)}$$

where

$$F(t) = \prod_{i=1}^{2g} (1 - \omega_i t).$$

The ω_i will be interpreted as the characteristic roots of π in a suitable representation in Chapter VII.

As an application of our computation of the number of rational points of C, let us take n large and relatively prime to any given integer. Then we see that there exists a prime rational cycle on C of degree relatively prime to any given integer, and hence there exists a cycle on C rational over k and of degree 1 (a fact originally due to F. K. Schmidt). From this, we get

PROPOSITION 3. *Let C be a complete non-singular curve, defined over a finite field k. Then there exists a canonical map $f: C \to J$ of C into its Jacobian, defined over k.*

Proof: Let $\mathfrak{a}$ be a cycle on C of degree 1, rational over k, and let $\mathfrak{a} = \sum n_i P_i$. Let P be a generic point of C over k. Let φ be as before a canonical admissible map of $C \times C$ into J defined over k. The point

$$\sum n_i \varphi(P, P_i)$$

is rational over $k(P)$ since $\mathfrak{a}$ is rational over k, and hence over

$k(P)$. We can write $\varphi(P, P_i) = f(P) - f(P_i)$ with some canonical map f of C into J, defined over k. Hence we find

$$\sum n_i \varphi(P, P_i) = f(P) - \sum n_i f(P_i)$$

taking into account the fact that deg $(\mathfrak{a}) = 1$. This shows that the rational map

$$P \rightarrow \sum n_i \varphi(P, P_i)$$

is a canonical map of C into J defined over k.

§ 4. *Reciprocity and correspondences*

We return to the theory of correspondences, and begin by giving some complements to the theorem of the cube and the square.

Let U, V be two complete varieties, non-singular in codimension 1. This insures that a divisor that is linearly equivalent to 0 has this property over a given field of rationality and also that such a divisor determines uniquely the function of which it is the divisor, up to a multiplicative constant.

For the rest of this section, a cycle of codimension 1 will be called a divisor. A cycle of dimension 0 will be called simply a cycle. Let $\mathfrak{a} = \sum n_i(P_i)$ be a cycle on U. Let g be a function on U such that g is defined at every point of $\mathfrak{a}$, and does not take on the value 0 at any of these points. Then we define

$$g(\mathfrak{a}) = \prod g(P_i)^{n_i}.$$

If $\mathfrak{a}$ is of degree 0, and h is a function such that $h = cg$ where c is constant, then $h(\mathfrak{a}) = g(\mathfrak{a})$.

Let D be a divisor on the product $U \times V$. Let $\mathfrak{a}$, $\mathfrak{b}$ be two cycles of degree 0 on U and V, respectively. Assume that $D(\mathfrak{a})$ is defined, and that it is the divisor of a function f on V. If in addition $f(\mathfrak{b})$ is defined, then we put

$$D(\mathfrak{a}, \mathfrak{b}) = f(\mathfrak{b})$$

and say that $D(\mathfrak{a}, \mathfrak{b})$ is defined.

Suppose that for every point P in $\mathfrak{a}$ and Q in $\mathfrak{b}$ the point (P, Q) does not lie in the support of D (or as we shall say more briefly, in D). In other words, suppose that no point of $\mathfrak{a} \times \mathfrak{b}$ is contained in D. Then first, $D(\mathfrak{a})$ is defined, and Q does not lie in the divisor of the function f. Hence f is defined at Q and $f(Q)$ is not equal to 0. Consequently $D(\mathfrak{a}, \mathfrak{b})$ is defined. This condition will be used constantly in the sequel. It is particularly useful, in view of the following lemma, whose proof I owe to Chevalley.

LEMMA 3. *Let V be an affine variety defined over a field k. Let D be a divisor on V, rational over k. Let S be a finite set of simple points of V. Then there exists a function φ on V, defined over k, such that no point of S lies in the support of $D + (\varphi)$.*

Proof: We may obviously assume that D is positive, and in fact a prime rational cycle. Furthermore, we may assume that the points of S are algebraic over k, because D contains the specializations over k of all of its points (IAG—III_5).

Let $R = k[V]$ be a coordinate ring for V over k, and let I be the ideal of R consisting of those functions f in R such that $(f) \geqq D$. This ideal has a finite basis, $I = (f_1, \ldots, f_n)$. Now let $\mathfrak{p}$ be a maximal ideal of R, and $\mathfrak{o} = R_{\mathfrak{p}}$ its local ring, with maximal ideal $\mathfrak{m}$. We assume $\mathfrak{p}$ belongs to a simple point, so that $\mathfrak{o}$ is a unique factorization domain, as in Appendix, § 2. Then there is a function t in $\mathfrak{o}$ which represents D locally at $\mathfrak{p}$. We can write $t = \sum z_i f_i$ with $z_i \in \mathfrak{o}$. I contend that if $w_i \in \mathfrak{o}$ is such that $w_i \equiv z_i \pmod{\mathfrak{m}}$ then $\sum w_i f_i$ differs from t by a unit in $\mathfrak{o}$. Indeed,

$$\sum w_i f_i = \sum (w_i - z_i) f_i + t.$$

Since $(f_i) \geqq D$, we can write $f_i = t g_i$ with $g_i \in \mathfrak{o}$. Hence $\sum w_i f_i$ is in $t(1 + \mathfrak{m})$, thereby proving our contention.

Now let $\mathfrak{p}_j$ be a finite number of maximal ideals of R, belonging to simple points. Let $\mathfrak{o}_j = R_{\mathfrak{p}_j}$ be their local rings with maximal ideals $\mathfrak{m}_j$. Given elements x_j in $\mathfrak{o}_j$, it is possible to find $x \in R$ such that $x \equiv x_j \pmod{\mathfrak{m}_j}$ by the well-known Chinese remainder theorem, applicable since $\mathfrak{p}_j + \mathfrak{p}_{j'} = R$ for $j \neq j'$. We now see that it suffices to approximate at each $\mathfrak{o}_j$ the coefficients of an element t_j representing D at $\mathfrak{o}_j$ in terms of the f_i. This proves our lemma.

If we wish to apply the local result to an abstract variety V, then we must assume that the given finite set of points (algebraic over k) can be represented on an affine k-open subset of V. This is the case on a projective variety.

Our goal is to prove Theorem 9 below, for abelian varieties. We must first deal with some general applications of the theorem of the square.

Let U, V, W, T be four varieties, defined over a field k. We assume them complete and non-singular in codimension 1.

Let D be a divisor on the product $U \times V \times W \times T$, rational over k. Let i, j, k, l range over 0 and 1, and take two copies of each one for our four varieties, U_i, V_j, W_k, T_l. On the double product

$$U_0 \times U_1 \times V_0 \times V_1 \times W_0 \times W_1 \times T_0 \times T_1 = U^{(2)} \times V^{(2)} \times W^{(2)} \times T^{(2)}$$

consider the divisor D_{ijkl} consisting of D on the partial product $U_i \times V_j \times W_k \times T_l$ taken with the full varieties on the others. (In other words, it is the inverse image of D under projection of the eightfold product to the fourfold product.) We can take a coboundary of D by setting

$$E = \sum (-1)^{i+j+k+l} D_{ijkl}.$$

Let

$$P = (u_0, u_1, v_0, v_1) \quad \text{and} \quad Q = (w_0, w_1, t_0, t_1)$$

be two independent generic points of $U^{(2)} \times V^{(2)}$ and $W^{(2)} \times T^{(2)}$ over k. By the theorem of the square, there exists a function f_P on $W \times T$, defined over $k(P)$, and a function g_Q on $U \times V$ defined over $k(Q)$ such that

$$(f_P) = \sum (-1)^{i+j} D(u_i, v_j) \quad \text{and} \quad (g_Q) = \sum (-1)^{k+l}\, {}^tD(w_k, t_l).$$

Furthermore, it is obvious from the definitions that we have

$$E(P) = \sum (-1)^{k+l} (f_P)_{kl} = \sum_{i,j,k,l,} (-1)^{i+j+k+l} D(u_i, v_j)_{kl},$$

$${}^tE(Q) = \sum (-1)^{i+j} (g_Q)_{ij} = \sum_{i,j,k,l,} (-1)^{i+j+k+l}\, {}^tD(w_k, t_l)_{ij}.$$

All further reciprocity theorems derived from the correspondence

D between $U \times V$ and $W \times T$ arise from the symmetry of the divisor E.

THEOREM 7. *The divisor E above is linearly equivalent to* 0.

Proof: There exists a function f^* on $U^{(2)} \times V^{(2)} \times W \times T$ defined over k such that

$$f^*(P, w, t) = f_P(w, t),$$

and we have

$$P \times (f_P) = (f^*) \cdot (P \times W \times T).$$

There exists a function F on the eightfold product such that

$$F(u_0, u_1, v_0, v_1, w_0, w_1, t_0, t_1) = \prod f^*(P, w_k, t_l)^{(-1)^{k+l}},$$

and from the definitions we see that

$$(F)(P) = E(P).$$

Hence (F) and E differ by a divisor which is degenerate on $U^{(2)} \times V^{(2)}$. Inducing (F) and E on the variety $U^{(2)} \times V^{(2)} \times W^{(2)} \times \Delta_T$ we obtain 0. Hence the degenerate component is equal to 0, and we have $(F) = E$, as desired.

Note that the function F such that $E = (F)$ is defined at the point obtained by setting $t_0 = t_1$, and takes the value 1. This gives us a way of normalizing it, since two functions representing E differ by a multiplicative constant.

On the other hand, from the symmetry of F, we see that if we go from right to left in the correspondence given by D, then we obtain the following reciprocity formula.

PROPOSITION 4. *Let F be a function on the eightfold product, defined over k, such that $E = (F)$, and normalized as above. Then with the above notation,*

$$F(P, Q) = \prod f_P(w_k, t_l)^{(-1)^{k+l}} = \prod g_Q(u_i, v_j)^{(-1)^{i+j}}.$$

As with the theorem of the square, we can formulate this result for special points. We need only proceed in the same manner as we did when we proved Theorem 3 of Chapter III, § 2 from the theorem of the cube, by repeating the above arguments.

THEOREM 8. *Let U, V, W, T be four varieties, complete and non-singular in codimension 1. Let (u'_0, u'_1, v'_0, v'_1) and (w'_0, w'_1, t'_0, t'_1) be two simple points of $U^{(2)} \times V^{(2)}$ and $W^{(2)} \times T^{(2)}$, respectively. Put*

$$\mathfrak{a} = \sum (-1)^{i+j} (u'_i, v'_j),$$
$$\mathfrak{b} = \sum (-1)^{k+l} (w'_k, t'_l).$$

Let D be a divisor on $U \times V \times W \times T$, and assume that none of the points (u'_i, v'_j, w'_k, t'_l) lies in D. Then

$$D(\mathfrak{a}, \mathfrak{b}) = {}^tD(\mathfrak{b}, \mathfrak{a}).$$

Proof: It is immediate from our hypothesis that the point (P', Q') (with an obvious notation) does not lie in $(F) = E$, and hence $F(P', Q')$ is defined. Our equality now follows by going through the symmetry arguments already carried out for generic points.

In the sequel, we shall not need any more the arguments preceding Theorem 8, only the statement of the theorem itself which we shall now apply to abelian varieties.

Let A, B be two abelian varieties and let D be a divisor on $A \times B$. Let $\mathfrak{a}$ be a cycle of dimension 0 and degree 0 on A, such that $D(\mathfrak{a})$ is defined. If $S(\mathfrak{a}) = 0$ then $D(\mathfrak{a})$ is linearly equivalent to 0 on B, by Corollary 2 of Theorem 4, Chapter III, § 3. Say $D(\mathfrak{a}) = (f)$. If now $\mathfrak{b}$ is a 0-cycle of degree 0 on B, also in the kernel of Albanese on B, i.e., such that $S(\mathfrak{b}) = 0$ on B, and such that no point of $\mathfrak{a} \times \mathfrak{b}$ is contained in D, then we may form symmetrically either $D(\mathfrak{a}, \mathfrak{b})$ or ${}^tD(\mathfrak{b}, \mathfrak{a})$.

THEOREM 9. *Let A, B be two abelian varieties. Let D be divisor on $A \times B$. Let $\mathfrak{a}$, $\mathfrak{b}$ be two cycles of dimension 0 and degree 0 on A, B, respectively, such that $S(\mathfrak{a}) = 0$ and $S(\mathfrak{b}) = 0$. Assume no point of $\mathfrak{a} \times \mathfrak{b}$ is contained in D. Then $D(\mathfrak{a}, \mathfrak{b})$ and ${}^tD(\mathfrak{b}, \mathfrak{a})$ are defined, and they are equal.*

Proof: Suppose first that D is the divisor of a function φ on $A \times B$. Our hypotheses imply that φ is defined at every point of $\mathfrak{a} \times \mathfrak{b}$, and our theorem simply asserts $\varphi(\mathfrak{a}, \mathfrak{b}) = \varphi(\mathfrak{a}, \mathfrak{b})$. Taking into account an obvious linearity, this remark allows us to change D by suitable linear equivalences, and to use Lemma 3.

We shall reduce our theorem to the case where $\mathfrak{a}$ and $\mathfrak{b}$ consist of four points each. This will then allow us to use Theorem 8.

Suppose that $\mathfrak{a}$ is written

$$\mathfrak{a} = (a_1) - (a_1') + (a_2) - (a_2') + \dots + (a_n) - (a_n').$$

Let x be any point of A. We can write

$$\begin{aligned}\mathfrak{a} = {} & (a_1) - (a_1') + (x) - (x + a_1 - a_1') + \\ & + (a_2) - (a_2') + (x + a_1 - a_1') - (x + a_2 - a_2' + a_1 - a_1') + \\ & + \dots\end{aligned}$$

so that we correct each successive step to cancel the preceding term. We must come back at the end to $-(x)$ using the fact that $S(\mathfrak{a}) = 0$. Thus our cycle can be written as a sum of cycles of degree 0, in the kernel of Albanese, and consisting of four points.

By the Lemma 3, we can move D by a linear equivalence so as to avoid all points in the product of the set of points entering in the expression for $\mathfrak{a}$ obtained above, with a similar set for $\mathfrak{b}$. In view of an obvious linearity, we see that it suffices to prove our theorem when $\mathfrak{a}$ and $\mathfrak{b}$ consist of four points.

Let us therefore write

$$\mathfrak{a} = (a_0) - (a_1) + (a_2) - (a_3),$$
$$\mathfrak{b} = (b_0) - (b_1) + (b_2) - (b_3).$$

Let $\lambda_A : A \times A \to A$ be the modified law of composition, such that $\lambda_A(u, v) = u - v$, and let $\lambda = (\lambda_A, \lambda_B)$. Let $D^* = \lambda^{-1}(D)$. It is a divisor on the product $A \times A \times B \times B$. Let u be a generic point of A, and let

$$u_0' = u + a_0, \quad u_1' = u + a_3, \quad v_0' = u, \quad v_1' = u + a_0 - a_1.$$

Do a similar construction for the points of $\mathfrak{b}$, to obtain points w_k' and t_l'. By hypothesis, no point of $\mathfrak{a} \times \mathfrak{b}$ lies in D, and hence none of the points (u_i', v_j', w_k', t') lies in D^*. If we put

$$\mathfrak{a}^* = \sum(-1)^{i+j}(u_i', v_j')$$

then

$$D^*(\mathfrak{a}^*) = \lambda_B^{-1} D(\mathfrak{a}).$$

This comes from commutativity in the diagram

$$\begin{array}{ccc} & (x,\ y)\times B\times B & \\ {}^{\text{inj}}\swarrow & & \searrow^{\lambda} \\ A\times A\times B\times B & & (x-y)\times B \\ {}_{\lambda}\searrow & & \swarrow_{\text{inj}} \\ & A\times B & \end{array}$$

where inj is the injection, together with Theorem 2 of Appendix, § 1, applied to the divisor D.

Similarly, we have a cycle $\mathfrak{b}^*$ such that

$$^tD^*(\mathfrak{b}^*) = \lambda_A^{-1}\,{}^tD(\mathfrak{b}).$$

If $(f) = D(\mathfrak{a})$, we have a function $f^* = f\lambda_B$ and if $(g) = {}^tD(\mathfrak{b})$ we have a function $g^* = g\lambda_A$. Then

$$(f^*) = \lambda_B^{-1}(f) = \lambda_B^{-1}D(\mathfrak{a}) \text{ and } (g^*) = \lambda_A^{-1}(g) = \lambda_A^{-1}\,{}^tD(\mathfrak{b}).$$

By Theorem 8, $f^*(\mathfrak{b}^*) = g^*(\mathfrak{a}^*)$ and hence clearly, $f(\mathfrak{b}) = g(\mathfrak{a})$. This proves our theorem.

We can now extend trivially our theorem to correspondences between arbitrary varieties.

THEOREM 10. *Let V, W be two complete varieties, non-singular in codimension* 1, *and such that any finite set of points can be represented on some affine open set. Let $\alpha : V \to A$ and $\beta : W \to B$ be canonical maps into their Albanese varieties. Let $\mathfrak{a}$, $\mathfrak{b}$ be two* 0-*cycles on V, W respectively, of degree* 0 *and such that $S(\alpha(\mathfrak{a})) = 0$ and $S(\beta(\mathfrak{b})) = 0$. Let D be a divisor on $V \times W$ such that no point of $\mathfrak{a} \times \mathfrak{b}$ lies in D. Then $D(\mathfrak{a}, \mathfrak{b})$ and ${}^tD(\mathfrak{b}, \mathfrak{a})$ are defined and they are equal.*

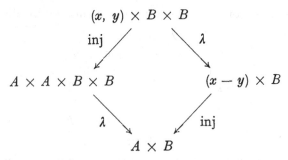

Proof: First observe that if our theorem is true for one divisor D in a correspondence class on $V \times W$ then it is true for every divisor D in that class not containing any point of $\mathfrak{a} \times \mathfrak{b}$. This is first checked for linear equivalence, and then for trivial divisors. A representative divisor in a correspondence class can always be obtained by an inverse image of a divisor on $A \times B$, by § 2, and hence we shall consider the following commutative diagram.

$$\begin{array}{ccc}
\mathfrak{a} \times W & \xrightarrow{(\alpha,\ \beta)} & \alpha(\mathfrak{a}) \times B \\
{\scriptstyle \text{inj}}\downarrow & & \downarrow{\scriptstyle \text{inj}} \\
V \times W & \xrightarrow[(\alpha,\ \beta)]{} & A \times B
\end{array}$$

Without loss of generality we may select a divisor D on $A \times B$ in the given correspondence class such that none of the points of $\alpha(\mathfrak{a}) \times \beta(\mathfrak{b})$ lies in D. If V, W are non-singular, we can apply the formalism of Appendix, § 1, but even if they have singularities, it is still true that in this special case, the composed inverse mapping of D is obtained by the two successive inverse maps (just as in § 2).

Let f be a function on B such that $D(\alpha(\mathfrak{a})) = (f)$. Then $D(\alpha(\mathfrak{a}), \beta(\mathfrak{b})) = f(\beta(\mathfrak{b}))$. On the other hand, there is a function f' on W such that $f' = f \circ \beta$, and we have $(f') = \beta^{-1}((f))$ (the middle variety is non-singular and we can apply Appendix, § 1, Theorem 1, Corollary 2). If we put $D' = (\alpha, \beta)^{-1}D$, we get

$$D'(\mathfrak{a}) = \beta^{-1}D(\alpha(\mathfrak{a})) = (f').$$

Hence $D'(\mathfrak{a}, \mathfrak{b}) = D(\alpha(\mathfrak{a}), \beta(\mathfrak{b}))$. Dealing in a similar manner with the transpose, we see that our theorem has become obvious.

Corollary. *Let C be a complete non-singular curve, and let $\mathfrak{a}$, $\mathfrak{b}$ be two divisors on C, linearly equivalent to* 0. *Let $\mathfrak{a} = (f)$ and $\mathfrak{b} = (g)$. Assume $\mathfrak{a}$, $\mathfrak{b}$ have no point in common. Then $f((g)) = g((f))$.*

Proof: Take the diagonal on $C \times C$. Our hypothesis means that no point of $\mathfrak{a} \times \mathfrak{b}$ is contained in the diagonal, and we can apply the theorem.

We can give an application of our reciprocity theorem to Kummer theory. Let V, W be as in Theorem 10. Let D be a divisor on the product. As before, let $\alpha : V \to A$ and $\beta : W \to B$ be two canonical maps into their Albanese varieties.

Let $\mathfrak{a}$, $\mathfrak{b}$ be two 0-cycles of degree 0 on A and B, respectively, and assume that there is an integer n such that $S(n\mathfrak{a}) = 0$ and $S(n\mathfrak{b}) = 0$. Let D_1 be linearly equivalent to D, and such that no point of $\mathfrak{a} \times \mathfrak{b}$ lies in D_1. Then one verifies immediately that

$$ {}^tD_1(n\mathfrak{b}, \mathfrak{a})/D_1(n\mathfrak{a}, \mathfrak{b}) $$

is an nth root of unity which is independent of the auxiliary divisor D_1 selected in the linear equivalence class of D. In fact, one also sees that D_1 could be selected in the correspondence class of D. We shall denote this root of unity by $e_{n,D}(\mathfrak{a}, \mathfrak{b})$. One also sees that it depends only on the equivalence class of $\mathfrak{a}$ and $\mathfrak{b}$ given by the sum, i.e., if $S(\alpha(\mathfrak{a}_1)) = S(\alpha(\mathfrak{a}))$ and $S(\beta(\mathfrak{b}_1)) = S(\beta(\mathfrak{b}))$ then $e_{n,D}(\mathfrak{a}, \mathfrak{b}) = e_{n,D}(\mathfrak{a}_1, \mathfrak{b}_1)$. Consequently, if $a = S(\alpha(\mathfrak{a}))$ and $b = S(\beta(\mathfrak{b}))$ are the associated points on the Albanese variety, then we can write

$$ e_{n,D}(a, b) = e_{n,D}(\mathfrak{a}, \mathfrak{b}). $$

As in the previous theorem, it is clear that our root of unity can be defined either by a correspondence on $V \times W$ or a correspondence on $A \times B$. We can state this formally in the following manner.

THEOREM 11. *Let V, W, A, B, α, β be as in the previous theorem. Let $\mathfrak{a}$, $\mathfrak{b}$ be two 0-cycles of degree 0 on V, W, respectively, such that $S(\alpha(n\mathfrak{a})) = 0$ and $S(\beta(n\mathfrak{b})) = 0$ for some integer n. Let D be a divisor on $A \times B$ which does not contain any point of $\alpha(\mathfrak{a}) \times \beta(\mathfrak{b})$. Let $D' = (\alpha, \beta)^{-1}D$. Then*

$$ e_{n,D}(\alpha(\mathfrak{a}), \beta(\mathfrak{b})) = e_{n,D'}(\mathfrak{a}, \mathfrak{b}). $$

Proof: The proof follows exactly the same pattern as the proof of the preceding theorem, and we leave it to the reader.

We return to abelian varieties. In Chapter VII we shall take another approach to our pairing, and we shall prove now a result which will establish the equivalence between them.

As we shall see in that chapter, if X is a divisor on A such that nX is linearly equivalent to 0, say $nX = (\varphi)$, then $(n\delta)^{-1}X = (\omega)$ for some function ω such that $\omega(u)^n = \varphi(nu)$. From this one sees immediately that $\omega(u + a)/\omega(u)$ is an nth root of unity for any point a on A such that $na = 0$. We shall also see trivially that this root of unity depends only on the linear equivalence class of X. The next theorem identifies it with $e_{n,D}(a, b)$. Note that in Chapter VII, we consider the case where $B = \hat{A}$ and D is a Poincaré divisor.

THEOREM 12. *Let D be a divisor on a product of abelian varieties $A \times B$. Let $\mathfrak{b}$ be a 0-cycle of degree 0 on B, such that $S(n\mathfrak{b}) = 0$. Assume ${}^tD(\mathfrak{b})$ is defined. Let ω be a function on A such that $(n\delta)^{-1}\, {}^tD(\mathfrak{b}) = (\omega)$. Let a be a point of A such that $na = 0$ and let u be a generic point of A. Let $b = S(\mathfrak{b})$. Then*

$$e_{n,D}(a, b) = \omega(a + u)/\omega(u).$$

Proof: Consider the rational map $n\delta : A \to A$ followed by the divisorial correspondence D on $A \times B$. Just as in the Appendix, § 2, we form the composed divisor $E = D \circ (n\delta)$ on $A \times B$. Then it is obvious that E and nD are in the same correspondence class on $A \times B$, and thus that

$$E = nD + (F) + X \times B + A \times Y$$

for some function F on $A \times B$, and divisors X on A and Y on B. Let k be a field of rationality for D, E, X, Y over which F is defined. We take two independent generic points u, w of A and B over k. Without loss of generality, we may assume that $\mathfrak{b} = (w + b) - (w)$. We put $\mathfrak{a} = (u + a) - (u)$. Then

$$\omega(u + a)/\omega(u) = {}^tD(\mathfrak{b}, \mathfrak{a}) = {}^tD(n\mathfrak{b}, \mathfrak{a})F(\mathfrak{a}, \mathfrak{b}).$$

On the other hand, $E(\mathfrak{a}) = D(nu + na) - D(nu) = 0$, and hence $E(\mathfrak{a}, \mathfrak{b}) = 1$. Thus $D(n\mathfrak{a}, \mathfrak{b})^{-1} = F(\mathfrak{a}, \mathfrak{b})$. This proves our theorem.

Historical note:

We have already said that the definition of the Picard variety taken here, and the canonical (and very simple and elegant) manner by which it is obtained from the Picard variety of the Albanese variety by means of the composition of divisors and the theory of correspondences (especially the seesaw principle) is due to Weil.

Of course, the Picard variety had already been constructed

previously by Matsusaka [53] and more recently by Chow [15] both of these authors making extensive use of the Jacobian of the generic curve. Matsusaka's construction is more explicit, while Chow's relies more on the universal mapping property, and thus Matsusaka's might be susceptible of giving a compatibility principle under specialization. Chow on the other hand also has the duality theory, and the fundamental fact that the Picard variety of V is the inverse image of the Picard variety of its Albanese variety. This is of importance for instance in the theory of coverings, where it is interpreted in terms of pull-back of coverings [43].

It cannot be said that the Picard variety is constucted in [69] because this paper begins by a false statement concerning the birational invariance. This is a delicate point, when the varieties involved have singularities. Chow makes a careful analysis of this situation, and I believe that he proves that if $f : U \to V$ is a birational correspondence between two varieties which are complete and non-singular in codimension 1, if $Y \in D_a(V)$ and if $f^{-1}(Y)$ is defined *and* in $D_a(U)$, then it belongs to ${}^t f_* \, [\mathrm{Cl}(Y)]$, i.e., to the transposed point on the Picard variety. In particular, if $Y \sim 0$ and if $f^{-1}(Y)$ is algebraically equivalent to 0, then it is linearly equivalent to 0 on U. We have not gone into these considerations here.

The theory of correspondences is of course very old, going back to Severi. It yields a satisfactory substitute for the representation of correspondences on the first homology group in the classical case. Our development is in large measure extracted from Weil's treatise [85], taking into account the complete reorganization of the theory as we have given it here, getting everything by canonical pull-backs.

For the arithmetic application, see the comments made at the end of the book, preceding the bibliography. We mention here merely that Weil was the first to perceive the connection between the diophantine analysis mod p and the Lefschetz fixed point formula [86]. For a discussion of the L-series of curves, see [84]. For the L-series of varieties, as well as an extension to arbitrary varieties of the analogue to Tchebotareff's density theorems in number theory, see [46], [47]. The local theory of L-series at ramified primes remains to be done in dimensions > 1. At the time this book is written, we do not yet have the conductor, for instance.

In this direction, we may give some indications concerning the possibility of defining generalized Picard and Albanese varieties. To begin with, Tate suggests the following definition of generalized Picard groups. One takes a finite set of simple points on the variety V. In each linear equivalence class, one can find a divisor which does not contain any of these points (say V is projective). Consider the functions on V which are defined at these points, taking the value 1, possibly with a high order of magnitude, determined by an ideal in the local ring of these points. The factor group of the divisors algebraically equivalent to 0 modulo the divisors of such functions is a generalized Picard group, or as one should probably say, an $\mathfrak{a}$-Picard group, $\mathfrak{a}$ being the ideal in the semilocal ring determined by the points.

One should then be able to show that these Picard groups can be parametrized by commutative algebraic groups, just like Rosenlicht's Jacobians, that is to say, that there exists a commutative group variety G and divisor on the product $V \times G$ satisfying the conditions of a Picard variety, where Cl now means the linear equivalence in the restricted sense.

As far as possible, one should work with non-complete varieties, perhaps by embedding them in complete ones, in order to apply the theory of correspondences to D on $V \times G$. Indeed, if $\mathfrak{a}$ is a 0-cycle of degree 0 on V, none of whose points is contained in a suitable divisor, then we can take $D(\mathfrak{a})$ on G, and then its associated point on the generalized Picard variety of G. This should give a canonical map of V into a generalized Albanese variety, and one can conjecture that the collection of all such maps obtained in the above manner satisfies the universal mapping property for maps of V into commutative groups, just as in [74].

The reciprocity theorem $f((g)) = g((f))$ on a curve arose first in Weil's investigations concerning class field theory [82] and has been recently used to deal with various questions of duality, for instance by Igusa [40] and Tate [81]. Its generalization to abelian varieties as it is formulated here was in fact motivated by the need to extend certain theorems from Jacobians to arbitrary abelian varieties, and by pull-back to varieties.

CHAPTER VII

The l-Adic Representations

In this chapter we exploit the fact that for n prime to the characteristic there exist exactly n^{2r} points of order n on an abelian variety A of dimension r.

To begin with, we obtain a representation of the ring of endomorphisms on a module of dimension $2r$ over the l-adic integers $\mathbf{Z}_l$, taking $n = l$ to be a prime number different from the characteristic. This allows us to show that End (A) is a module of finite type over the integers $\mathbf{Z}$, and hence that the algebra of endomorhpisms $\text{End}_{\mathbf{Q}}$ (A) is finite dimensional over $\mathbf{Q}$, and thus semisimple. We shall also identify the characteristic polynomial of an endomorphism with the characteristic polynomial of the matrix associated with this endomorphism in the l-adic representation. From this we deduce that the Néron-Severi group $N'(A)$ is of finite type, taking into account the fact that $X \to \varphi_X$ induces an isomorphism of $N'(A)$ into $H(A, \hat{A})$.

Next, we show how the l-adic spaces of A and $\hat{A}$ are in duality with each other by means of Kummer theory, and how the transpose of a homomorphism corresponds to the transpose of the induced linear transformation on the l-adic spaces. The l-adic statements occur as trivial limiting consequences of duality statements concerning the points of order n on A and $\hat{A}$.

These duality statements could be obtained from another point of view, as in the last chapter, where we obtained a general reciprocity formula arising from a divisorial correspondence between two abelian varieties, one of whose applications is to Kummer theory.

§ 1. *The l-adic spaces*

In order to construct the l-adic representations, we are going to use a canonical construction due to Tate. Let G be an additive abelian group which is infinitely divisible. We associate with G the *Tate group* $T(G)$. It is the set of all vectors $(a_n)_{n \in \mathbf{Z}}$ with $n > 0$ with components $a_n \in G$ satisfying the condition $m \cdot a_{mn} = a_n$. Addition is defined componentwise.

For our purposes, we prefer to restrict ourselves to that portion of the Tate group belonging to a fixed prime number l. For this, we need only assume that G is infinitely divisible by powers of l, i.e., that $lG = G$. The group $T_l(G)$ consists of vectors

$$(a_1, a_2, \ldots, a_n, \ldots)$$

with denumerably many components $a_i \in G$, satisfying $la_{i+1} = a_i$ and $la_1 = 0$. The notation is thus different from that of the first paragraph, and it has been more convenient to take the logarithm to the base l for the indices, in order to avoid double indices. Addition in $T_l(G)$ is defined componentwise. We can define on $T_l(G)$ the structure of a module over the l-adic integers $\mathbf{Z}_l$ as follows. Let $z \in \mathbf{Z}_l$. Let m be an integer such that $m \equiv z \pmod{l^n}$. If $l^n a = 0$, we define $za = ma$, this being independent of the auxiliary m selected subject to the above condition. We then define the operation of z on $T_l(G)$ by

$$z \cdot (a_1, a_2, \ldots) = (za_1, za_2, \ldots).$$

It is clear that this definition satisfies the usual conditions for one group to operate on another.

We shall take for G the group of points on A whose order is a power of the prime number l, denoted by $\mathfrak{g}_l(A)$. We write $T_l(A)$ instead of $T_l(\mathfrak{g}_l(A))$. We shall from now on assume that $l \neq p$.

The finite group of points $a \in A$ such that $l^n a = 0$ will be denoted by $\mathfrak{g}_n(A)$ (the prime number l having been fixed). The nth component of a vector of $T_l(A)$ is therefore in $\mathfrak{g}_n(A)$.

The group $\mathfrak{g}_1(A)$ is a vector space over the prime field $\mathbf{Z}/l\mathbf{Z}$. The theorem according to which there exist l^{2r} points of order l on A means that the dimension of this space is equal to $2r$.

(If we had built up the group $T_p(A)$, the dimension would have been smaller than $2r$.)

Note that $T_l(A)$ is obviously without torsion over $\mathbf{Z}_l$. We are going to see that it is a module of finite type of dimension $2r$ over $\mathbf{Z}_l$. (We may speak of dimension in view of the structure of $\mathbf{Z}_l$.) Indeed, let $x_1, \ldots, x_{2r}$ be vectors of $T_l(A)$ whose first components $a_{1,1}, \ldots, a_{2r,1}$ are linearly independent over the field $\mathbf{Z}/l\mathbf{Z}$. Then these vectors are linearly independent over $\mathbf{Z}_l$; for, if we had a relation of linear dependence, we could assume that not all the coefficients are divisible by l, and hence the projection of this relation on the first component would contradict the hypothesis made on the a_{ij}.

I contend that the x_i form a basis of $T_l(A)$ over $\mathbf{Z}_l$. We are going to prove this by an inductive argument. Suppose that we can write every element w of $T_l(A)$ as a linear combination

$$w \equiv z_1 x_1 + \ldots + z_{2r} x_{2r} \pmod{l^n T_l(A)} \tag{1}$$

with integers $z_j \in \mathbf{Z}$.

Let $w = (b_1, \ldots, b_n, b_{n+1}, \ldots)$. By definition, we have for the first $n + 1$ components,

$$z_1(a_{1,1}, \ldots, a_{1,n+1}) + \ldots + z_{2r}(a_{2r,1}, \ldots, a_{2r,n+1})$$
$$= (b_1, \ldots, b_n, b_{n+1}) + (0, \ldots, 0, c_{n+1})$$

for some $c_{n+1} \in \mathfrak{g}_{n+1}(A)$. By the very choice of the vectors x_i, there exist integers $d_1, \ldots, d_{2r}$ such that

$$c_{n+1} = d_1 l^n a_{1,n+1} + \ldots + d_{2r} l^n a_{2r,n+1}.$$

If we replace $z_1, \ldots, z_{2r}$ by $z_1 + d_1 l^n, \ldots, z_{2r} + d_{2r} l^n$, we see that we have extended the congruence (1) from n to $n + 1$. This gives us what we wanted. Summarizing, we get:

PROPOSITION 1. *The module $T_l(A)$ over $\mathbf{Z}_l$ is of dimension $2r$.*

It will sometimes be useful to work with a vector space over the l-adic numbers $\mathbf{Q}_l$ instead of a module over the l-adic integers. For this purpose, we define the *extended Tate group* $E_l(A)$ in the following manner. It is the group of vectors

$$(a_0, a_1, a_2, \ldots)$$

with $a_i \in \mathfrak{g}_i(A)$, satisfying $la_i = a_{i-1}$, but this time with a_0 arbitrary. We have therefore an exact sequence

$$0 \to T_l(A) \to E_l(A) \to \mathfrak{g}_l(A) \to 0$$

obtained by taking the projection on the component a_0 of an element of $E_l(A)$. As before, one sees that $E_l(A)$ is a module over $\mathbf{Z}_l$. The map

$$x \to lx$$

of $E_l(A)$ into itself is in fact surjective, because

$$l(a_1, a_2, \ldots) = (a_0, a_1, a_2, \ldots).$$

One sees immediately that it is injective, i.e., that $lx = 0$ implies $x = 0$. The group $E_l(A)$ is therefore infinitely and uniquely divisible by powers of l. Since every l-adic number can be written $l^m z$ with $z \in \mathbf{Z}_l$ and $m \in \mathbf{Z}$, we see immediately that $E_l(A)$ is a vector space over $\mathbf{Q}_l$. It is obviously isomorphic with the tensor product

$$T_l(A) \otimes \mathbf{Q}_l$$

taken over $\mathbf{Z}_l$, and is of dimension $2r$ over $\mathbf{Q}_l$.

Let $\alpha \in H(A, B)$ be a homomorphism. Then α induces a $\mathbf{Z}_l$-linear map

$$\alpha_T : T_l(A) \to T_l(B)$$

by the formula

$$\alpha_T(a_1, a_2, \ldots) = (\alpha a_1, \alpha a_2, \ldots).$$

The $\mathbf{Z}$-module $H(A, B)$ becomes a module over the l-adic integers $\mathbf{Z}_l$ if we define $(z\alpha)x = \alpha(zx)$ for $x \in T_l(A)$ and $z \in \mathbf{Z}_l$. We obtain therefore a representation of $H(A, B)$. Similarly, we get a representation of $H_{\mathbf{Q}}(A, B)$, whose elements induce $\mathbf{Q}_l$-linear transformations of $E_l(A)$ into $E_l(B)$. For $\alpha \in H_{\mathbf{Q}}(A, B)$, we may denote by α_E the induced transformation on $E_l(A)$. If A and B have the same dimension, and if $\nu(\alpha) \neq 0$, then α_E is surjective because the kernel of α is finite. We note that in general, α_T is not surjective. More precisely, we have the following result.

THEOREM 1. *Let $\alpha : A^r \to B$ be a homomorphism. In order that the kernel of α be finite, it is necessary and sufficient that the linear* *transformation* α_T *be of rank* $2r$. *Suppose in addition that* $\nu(\alpha) \neq 0$ *so that* $\dim A = \dim B = r$. *Then the highest power of* l *which divides* $\nu(\alpha)$ *is equal to the highest power of* l *which divides the determinant* $\det(\alpha_T)$.

Proof: The first assertion is an immediate consequence of the definitions. As to the second, we use the theory of elementary divisors, which allows us to select a basis for $T_l(A)$ over $\mathbf{Z}_l$ and a basis of $T_l(B)$ over $\mathbf{Z}_l$ such that the matrix representing α_T is of type

$$\begin{pmatrix} l^{m_1} & & & & 0 \\ & \cdot & & & \\ & & \cdot & & \\ & & & \cdot & \\ 0 & & & & l^{m_{2r}} \end{pmatrix}.$$

The definition of the Tate vectors shows immediately that the kernel of α is exactly divisible by the product $l^{m_1} \ldots l^{m_{2r}}$.

One sees immediately that $\nu(\alpha) = 0$ in the above theorem if and only if $\det(\alpha_T) = 0$.

We are going to show that the representation of $H(A, B)$ in our l-adic module corresponds to the tensor product of $H(A, B)$ with $\mathbf{Z}_l$. We axiomatize our arguments.

THEOREM 2. *Let* R *be an algebra over the integers* $\mathbf{Z}$ *(not necessarily of finite type). Suppose that there exists a symmetric bilinear form* (α, β) *on* R *with values in* $\mathbf{Z}$, *such that* $(\alpha, \alpha) > 0$ *if* $\alpha \neq 0$. *Let* $\alpha \to M(\alpha)$ *be a representation of* R *into the ring of* $\mathbf{Z}_l$*-linear transformations of a module* V *over* $\mathbf{Z}_l$. *Suppose in addition that if* $\beta \in R$ *and if* $M(\beta)V \subset l^n V$ *then there exists an element* $\beta_1 \in R$ *such that* $\beta = l^n \beta_1$. *Then if* α_i $(i = 1, \ldots, d)$ *are elements of* R *linearly independent over* $\mathbf{Z}$, *then their representations* $M(\alpha_i)$ *are linearly independent over* $\mathbf{Z}_l$.

Proof: We may assume without loss of generality that the α_i are orthogonal, i.e., that $(\alpha_i, \alpha_j) = 0$ of $i \neq j$. Suppose that there exists a relation

$$\sum z_i M(\alpha_i) = 0$$

with $z_i \in \mathbf{Z}_l$, $z_1 \neq 0$. Let $z_i^{(n)}$ be elements of $\mathbf{Z}$ such that

$$z_i^{(n)} \equiv z_i \pmod{l^n}$$

with n large. Put $\beta = \sum z_i^{(n)} \alpha_i$. Then $M(\beta)V \subset l^n V$, and by hypothesis, there exists $\beta_1 \in R$ such that $\beta = l^n \beta_1$. We find

$$l^n(\beta_1, \alpha_1) = (\beta, \alpha_1) = z_1^{(n)}(\alpha_1, \alpha_1).$$

For m large, we see that l^m divides z_1, a contradiction.

The hypotheses of Theorem 2 are applicable to End (A), because even though the bilinear form of Chapter V, § 3 takes on its values in $\mathbf{Q}$, the denominators are uniformly bounded by deg (ξ^r). Multiplying this form by this integer, we find a bilinear form of the type considered in Theorem 2. In addition, the divisibility condition on β is satisfied, because the hypothesis $M(\beta)V \subset l^n V$ in the special case under consideration implies that the kernel of β contains the points of order l^n on A. Since the homomorphism $l^n \delta_A$ satisfies the universal mapping property for every homomorphism whose kernel contains these points, we see that $\beta = l^n \beta_1$ for some $\beta_1 \in$ End (A).

Corollary 1. *The hypotheses being the same as in Theorem 2, if V is of finite type over $\mathbf{Z}_l$, then R is a module of finite type over $\mathbf{Z}$.*

Proof: Let $\alpha_1, \ldots, \alpha_d$ be a maximal system of elements of R whose representations $M(\alpha_i)$ are linearly independent over $\mathbf{Z}_l$. We may assume that the α_i are orthogonal. Let α be an arbitrary element of R. Let D be the product of the (α_i, α_i). It is a positive integer. By the theorem, and our hypothesis on the α_i, we can write

$$Dm \cdot \alpha = m_1 \alpha_1 + \ldots + m_d a_d$$

with suitable integers m, m_i, and we must have $m \neq 0$. This gives $Dm(\alpha, \alpha_i) = m_i(\alpha_i, \alpha_i)$, and consequently m divides each m_i. Hence $D \cdot \alpha$ is in the module generated by the α_i. The mapping $\alpha \to D \cdot \alpha$ is an additive isomorphism of R into a finitely generated group, and hence R is of finite type.

We express our results for the special case $R =$ End (A).

Corollary 2. *The group $H(A, B)$ is of finite type. If $\alpha_1, \ldots, \alpha_d$*

are elements of $H(A, B)$ *which are linearly independent over* $\mathbf{Z}$, *then they are independent over* $\mathbf{Z}_l$. *The map* $\alpha \to \alpha_T$ *of* End (A) *into the ring of* $\mathbf{Z}_l$*-linear transformations of* $T_l(A)$ *is a ring isomorphism, and* End (A) *is a module of finite type, without torsion, of dimension* $\leqq 4r^2$.

COROLLARY 3. *The group* $N'(A)$ *is of finite type.*

Proof: The map $X \to \varphi_X$ where X ranges over the divisors of A induces an isomorphism of $N'(A)$ into $H(A, \hat{A})$, which is of finite type.

COROLLARY 4. *The algebra* $\mathrm{End}_{\mathbf{Q}}(A)$ *is semi-simple.*

Proof: This is immediate from the results already obtained.

In order to show that the characteristic polynomial of an endomorphism coincides with the characteristic polynomial of its representation in $T_l(A)$, we need a lemma.

LEMMA 1. *Let*

$$P(X) = X^r + \sum_{i=1}^{r} a_i X^{r-i},$$

$$Q(X) = X^r + \sum_{i=1}^{r} b_i X^{r-i}$$

be two polynomials with coefficients in $\mathbf{Q}_l$. *Let* K *be a finite algebraic extension of* $\mathbf{Q}_l$ *in which* $P(X)$ *and* $Q(X)$ *split into linear factors,* $P(X) = \prod (X-\omega_i)$ *and* $Q(X) = \prod (X - \eta_j)$. *Suppose that for every polynomial* F *with coefficients in* $\mathbf{Z}$, *we have*

$$\|\prod F(\omega_i)\| = \|\prod F(\eta_j)\|$$

where $\|\ \|$ *is the* l*-adic absolute value. Then* $P(X) = Q(X)$.

Proof: Select one of the roots of P or Q, say ω_1. It will obviously suffice to prove that the multiplicity d of ω_1 in P is equal to its multiplicity e in Q. In the statement of the lemma, we took F with coefficients in $\mathbf{Z}$. By continuity, our hypothesis continues to hold for coefficients in $\mathbf{Z}_l$ and by homogeneity, for coefficients in $\mathbf{Q}_l$. Let β be an l-adic number which is very close to ω_1, but $\beta \neq \omega_1$. Then clearly,

$$\prod_i \|(\beta - \omega_i)\| = \|\beta - \omega_1\|^d \prod_{\omega_i \neq \omega_1} \|\omega_1 - \omega_i\|$$

and

$$\prod_j \|(\beta - \eta_j)\| = \|\beta - \omega_1\|^e \prod_{\eta_j \neq \omega_1} \|\omega_1 - \eta_j\|.$$

Let m be the degree of β over $\mathbf{Q}_l$, and let F be the irreducible polynomial of β over $\mathbf{Q}_l$. Let (σ) range over distinct automorphisms of the algebraic closure of $\mathbf{Q}_l$, such that their restrictions to $\mathbf{Q}_l(\beta)$ give all distinct conjugates of β. Note that each σ permutes the ω_i and the η_j. In view of the known elementary fact that the absolute value of a conjugate of an element of $\overline{\mathbf{Q}}_l$ is equal to the absolute value of that element, we get:

$$\begin{aligned} \|\prod_i (\beta - \omega_i)\|^m &= \|\prod_\sigma \prod_i (\beta - \omega_i)\| \\ &= \|\prod_\sigma \prod_i (\beta^\sigma - \omega_i^\sigma)\| \\ &= \|\prod_\sigma \prod_i (\beta^\sigma - \omega_i)\| \\ &= \|\prod_i F(\omega_i)\|. \end{aligned}$$

Similarly, we get

$$\|\prod_j (\beta - \eta_j)\|^m = \|\prod_j F(\eta_j)\|.$$

In view of our hypothesis, and of the two equalities obtained above, we see that

$$\|\beta - \omega_1\|^d \prod_{\omega_i \neq \omega_1} \|\omega_1 - \omega_i\| = \|\beta - \omega_1\|^e \prod_{\eta_j \neq \omega_1} \|\omega_1 - \eta_j\|.$$

Since we select β very close to ω_1, we see that such an equality can hold only if $d = e$, because the expressions not involving β remain constant. This proves our lemma.

THEOREM 3. *Let A^r be an abelian variety and α an endomorphism of A. Then the characteristic polynomial of α is equal to the characteristic polynomial of the linear transformation α_T induced by α on $T_l(A)$. In particular, we have*

$$\nu(\alpha) = \det(\alpha_T),$$

and $\operatorname{tr}(\alpha)$ *is equal to the trace of* α *in this representation.*

Proof: The assertion concerning the characteristic polynomial comes from the lemma, from Theorem 8 of Chapter IV, § 3, and from Theorem 1.

COROLLARY 1. *Let* A, B *be two abelian varieties, and let* $\alpha \in H_{\mathbf{Q}}(A, B)$, $\beta \in H_{\mathbf{Q}}(B, A)$. *Then* $\operatorname{tr}(\alpha\beta) = \operatorname{tr}(\beta\alpha)$.

Proof: This comes from the elementary theory of matrices.

COROLLARY 2. *If* $\alpha \in \operatorname{End}(A)$, *then the coefficients of the characteristic polynomial are rational integers.*

Proof: We have seen that $\operatorname{End}_{\mathbf{Q}}(A)$ is semi-simple, and that $\operatorname{End}(A)$ is of finite type, without torsion over $\mathbf{Z}$. Every element of $\operatorname{End}(A)$ is therefore integral over $\mathbf{Z}$. The minimal polynomial of each element $\alpha \in \operatorname{End}(A)$ has integral coefficients and its roots are algebraic integers. Our corollary is therefore a consequence of the theory of representations.

§ 2. *Dual representations*

Let A be an abelian variety and $\hat{A}$ its Picard variety. We are going to show that the $\mathbf{Z}_l$-modules $T_l(A)$ and $T_l(\hat{A})$ are dual over $\mathbf{Z}_l$ in the ordinary sense of the word. We continue to assume that l is a prime number unequal to the characteristic, but first work with an arbitrary positive integer n.

The mapping $n\delta_A$ on A gives a Kummer covering of A if n is prime to p, and we are interested in seeing what happens to divisor classes for linear equivalence under their inverse image in this covering.

We may work with n possibly divisible by p, although in this case, our considerations are going to be trivial. Let X be a divisor on A, and assume that $X \equiv 0$ and $nX \sim 0$. According to the corollary of Theorem 2, Chapter IV, § 1 we get

$$(n\delta)^{-1}(X) \sim nX \sim 0.$$

Hence there exists a function ω on A such that $(n\delta)^{-1}(X) = (\omega)$. Furthermore, if k is a field of definition for A over which X is rational, then we may assume that ω is defined over k since

$(n\delta)^{-1}(X)$ is also rational over k. Put $nX = (\varphi)$ where φ is a function on A, defined over k, and consider the function θ on A such that $\theta(u) = \varphi(nu)$, so that θ is the composed function $\varphi(n\delta)$. Applying Theorem 2 of the Appendix, § 1 we find

$$\begin{aligned}(\theta) &= \theta^{-1}[(0)-(\infty)]\\ &= (n\delta)^{-1}\{\varphi^{-1}[(0)-(\infty)]\}\\ &= (n\delta)^{-1}((\varphi))\\ &= (n\delta)^{-1}(nX)\\ &= n\cdot(\omega)\\ &= (\omega^n).\end{aligned}$$

It follows that there exists a constant c such that $\theta = \omega^n c$, and c is rational over k since θ and ω are defined over k. We could have selected ω such that

$$\varphi(nu) = \omega(u)^n.$$

In short, we have:

PROPOSITION 2. *Let A be defined over k, and let X be a divisor on A, rational over k such that*

$$nX = (\varphi)$$

with a function φ defined over k. Then there exists a function ω on A, defined over k, such that $(n\delta)^{-1}(X) = (\omega)$ and we have

$$\varphi(nu) = \omega(u)^n$$

for u generic. Every function ω satisfying this relation is uniquely determined up to a root of unity.

Since we work at the moment with an arbitrary n, we do not use the logarithmic indices of § 1, and denote for a while by $\mathfrak{g}_n(A)$ the group of points $a \in A$ such that $na = 0$. We denote by $\hat{\mathfrak{g}}_n(A)$ the divisor classes ξ of $D(A)/D_l(A)$ such that $n\xi = 0$. Later, we shall consider the group $\mathfrak{g}_n(\hat{A})$ which can be viewed in an obvious fashion as a subgroup of $\hat{\mathfrak{g}}_n(A)$. To say that there is no torsion on A amounts to saying that these two groups are equal (by Corollary 4 of Theorem 4, Chapter IV, § 2).

PROPOSITION 3. *The notations being the same as in Proposition* 2, *let* a *be a point of* A *such that* $na = 0$. *Then we can write*

$$\omega(u + a) = e_n(a, X)\omega(u)$$

with some nth *root of unity* $e_n(a, X)$. *If* $X' \sim X$, *then* $e_n(a, X') = e_n(a, X)$ *in other words, this root of unity depends only on the linear equivalence class* $\mathrm{Cl}(X)$.

Proof: Write $X' = X + (f)$ with some function f. Then $nX' = (\varphi')$ with $\varphi' = \varphi f^n$, and $\varphi'(nu) = \omega'(u)^n = \varphi(nu)f(nu)^n$. We can therefore take ω' such that $\omega'(u) = \omega(u)f(nu)$. This shows that

$$\frac{\omega'(u + a)}{\omega'(u)} = \frac{\omega(u + a)f(nu)}{\omega(u)f(nu)} = e_n(a, X).$$

If we denote by ξ the linear equivalence class of X, we see immediately that the map

$$(a, \xi) \to e_n(a, \xi)$$

with $a \in \mathfrak{g}_n(A)$ and $\xi \in \hat{\mathfrak{g}}_n(A)$ is a bilinear map of these two groups into the group of nth roots of unity. The purpose of the following proposition is to determine the kernels.

PROPOSITION 4. *Let* n *be prime to* p. *If* $e_n(a, \xi) = 1$ *for all* $\xi \in \hat{\mathfrak{g}}_n(A)$ *and* a *fixed, then* $a = 0$. *If* $e_n(a, \xi) = 1$ *for all* $a \in \mathfrak{g}_n(A)$ *and* ξ *fixed, then* $\xi = 0$. *The groups* $\mathfrak{g}_n$ *and* $\hat{\mathfrak{g}}_n$ *are each other's character groups, put in duality by the bilinear form* $e_n(a, \xi)$.

Proof: Let us first show that if $e_n(a, \xi) = 1$ for all a and ξ fixed, then $\xi = 0$. Our hypothesis means that $\omega(u + a) = \omega(u)$ for all $a \in \mathfrak{g}_n$. We may assume the ground field k so large that all the points of $\mathfrak{g}_n$ are rational over k. The function ω in the field $k(A)$ is then invariant by the translations of $\mathfrak{g}_n$. The field of invariants by this group of translations is $k(nA)$, i.e., we can find a function f on A such that $\omega(u) = f(nu)$. We can also take f defined over k. This gives $\varphi(nu) = f(nu)^n$, and hence $\varphi(u) = f(u)^n$. Let Y be the divisor of f. We have $nY = (\varphi) = nX$, and hence $X = Y$. We see that X is linearly equivalent to 0, and thus $\xi = 0$.

The elementary duality theory of finite abelian groups shows that if $\mathfrak{h}$ denotes the kernel of $\mathfrak{g}_n$, i.e., the set of points $a \in \mathfrak{g}_n$ such that $e_n(a, \xi) = 0$ for all $\xi \in \hat{\mathfrak{g}}_n$, then $\mathfrak{g}_n/\mathfrak{h}$ is in exact duality with $\hat{\mathfrak{g}}_n$. In particular, the order of the two groups $\mathfrak{g}_n/\mathfrak{h}$ and $\hat{\mathfrak{g}}_n$ is the same. Since $\hat{\mathfrak{g}}_n$ contains the subgroup of points of $\mathfrak{g}_n(\hat{A})$, and since the Picard variety $\hat{A}$ is isogenous to A, we know that the order of $\hat{\mathfrak{g}}_n$ is $\geqq n^{2r}$. But the order of $\mathfrak{g}_n$ is exactly n^{2r}. Hence $\mathfrak{h} = 0$, and $\hat{\mathfrak{g}}_n(A) = \mathfrak{g}_n(\hat{A})$ is the group of points of order n on the Picard variety. This proves our proposition, and also takes care of the torsion prime to p on an abelian variety.

COROLLARY. *The period of every element of the torsion group $D_\tau(A)/D_a(A)$ is a power of p. If A is a Jacobian, then this group is trivial.*

Proof: We use again Corollary 4 of Theorem 4, Chapter IV, § 2. On a Jacobian, we had given a special argument to show that there is no p-torsion (Chapter VI, § 3, Proposition 2).

We shall eventually pass to the limit, and for this purpose we give a consistency formula.

PROPOSITION 5. *Let $a \in \mathfrak{g}_n(A)$ and $\xi \in \hat{\mathfrak{g}}_n(A)$. Suppose that $mna = 0$ and $n\xi = 0$. Then*

$$e_{mn}(a, \xi) = e_n(ma, \xi).$$

If $mna = 0$ and $mn\xi = 0$, then

$$e_{mn}(a, \xi)^m = e_n(ma, m\xi).$$

Proof: Let $nX = (\varphi)$ and $(n\delta)^{-1}(X) = (\omega)$ as before. We have $\omega(u + ma) = e_n(ma, X)\omega(u)$, and $mnX = (\varphi^m)$. Since $\omega(u)^n = \varphi(nu)$, we find $\omega(mu)^n = \varphi(mnu)$ and consequently $\omega(mu)^{mn} = \varphi(mnu)^m$. This means that if we take the function ω' defined by $\omega'(u) = \omega(mu)$, and if φ' is the function defined by $mnX = (\varphi')$, then $\omega'(u + a) = e_{mn}(a, X)\omega'(u)$ by definition. On the other hand,

$$\omega'(u + a) = \omega(mu + ma) = e_n(ma, X)\omega(mu),$$
$$\omega'(u) = \omega(mu).$$

The first assertion is then a consequence of the definitions. The second is a consequence of the first, and of the relation

$$e_{mn}(a, \xi)^m = e_{mn}(a, m\xi)$$

which is true by linearity.

PROPOSITION 6. *Let $\alpha : A \to B$ be a homomorphism, and Y a divisor on B such that $nY \sim 0$. Then if $a \in \mathfrak{g}_n(A)$, we have*

$$e_n(\alpha a, Y) = e_n(a, \alpha^{-1}(Y))$$

whenever $\alpha^{-1}(Y)$ is defined.

Proof: Put $(n\delta)^{-1}(Y) = (\omega)$. By definition, for v generic on B, we have

$$\omega(v + \alpha a) = \omega(v) e_n(\alpha a, Y).$$

Put $X = \alpha^{-1}(Y)$. Then

$$\begin{aligned}(n\delta)^{-1}(X) &= \alpha^{-1}(n\delta)^{-1}(Y)\\ &= \alpha^{-1}[(\omega)].\end{aligned}$$

Consequently the function $\omega \circ \alpha$ on A is such that $(n\delta)^{-1}(X) = (\omega \circ \alpha)$. If u is a generic point of A such that $\alpha u = v$, we see that our formula is now a consequence of the definitions.

The above proposition will show later that the representations of α and ${}^t\alpha$ on the l-adic modules are transpose to each other. The next proposition gives us a interesting special case of our symbol $e_n(a, \xi)$.

PROPOSITION 7. *Let $a \in \mathfrak{g}_n(A)$ and let X be an arbitrary divisor on A. Then*

$$e_n(a, X_a - X) = 1.$$

Proof: Put $(\varphi) = n(X_a - X)$ and $\omega(u)^n = \varphi(nu)$ as usual. We have to show that $\omega(u - a) = \omega(u)$. We can find a point b of A such that $a = nb$. Consider the function

$$\psi(u) = \prod_{i=0}^{n-1} \omega(u - ib).$$

We have

$$\psi(u)^n = \prod_{i=0}^{n-1} \varphi(nu - inb) = \prod_{i=0}^{n-1} \varphi(nu - ia).$$

Put $\theta(u) = \prod_{i=0}^{n-1} \varphi(u - ia)$. Since $n(X_a - X) = (\varphi)$, we get

$$(\theta) = n \sum_{i=0}^{n-1} [X_{a+ia} - X_{ia}] = 0.$$

Indeed, recall that if we denote by φ_c the function such that $\varphi_c(u) = \varphi(u - c)$, then $(\varphi_c) = n(X_{a+c} - X_c)$. It follows that θ is constant, and ψ is therefore constant. We conclude that $\psi(u - b) = \psi(u)$, and by the definition of ψ, we get

$$\prod_{i=0}^{n-1} \omega(u - ib) = \prod_{i=1}^{n} \omega(u - ib)$$

whence $\omega(u) = \omega(u - nb) = \omega(u - a)$. This proves the proposition.

We shall express all the preceding results in a convenient language, in terms of the Tate group. We still have l prime to the characteristic, and return to the notation where $\mathfrak{g}_n(A)$ denotes the group of points $a \in A$ such that $l^n a = 0$. Let $x \in T_l(A)$ and $\hat{y} \in T_l(\hat{A})$. We can identify the additive group $\mathbf{Z}_l$ of l-adic integers with the Tate group associated with the group of l^νth roots of unity, which is infinitely divisible by powers of l. Put $x = (a_1, a_2, \ldots)$ and $\hat{y} = (\eta_1, \eta_2, \ldots)$. For a_n, η_n we know how to define $e_n(a_n, \eta_n)$ by means of Proposition 3. In addition, Proposition 5 shows that the vector

$$(e_1(a_1, \eta_1), e_2(a_2, \eta_2), \ldots)$$

is an element of the Tate group associated with the l^νth roots of unity, identified with $\mathbf{Z}_l$. If we denote this vector by $(x, \hat{y})$ then we see immediately that we have defined a bilinear form on $T_l(A) \times T_l(\hat{A})$ with values in $\mathbf{Z}_l$. It is also clearly $\mathbf{Z}_l$-linear. We shall see in a moment that it is skew-symmetric.

The next two theorems give the main properties of this form.

THEOREM 4. *Let $\alpha : A \to B$ be a homomorphism, $x \in T_l(A)$ and $\hat{y} \in T_l(\hat{B})$. Then we have*

$$(\alpha_T x, \hat{y}) = (x, {}^t\alpha_T \hat{y}).$$

Proof: This is a translation of Proposition 6.

Of course, the transpose ${}^t\alpha_T$ must be read $({}^t\alpha)_T$. The theorem expresses the fact that the transpose of α is represented in the Tate module by the transpose of the linear transformation. As a corollary, we have:

COROLLARY. *Let α be an endomorphism of A. Then $\nu(\alpha) = \nu({}^t\alpha)$.*

Proof: It suffices to use Theorem 3 of § 1.

To simplify the notation in the next theorem, we allow ourselves to write α instead of α_T for the induced representation of α in $T_l(A)$.

THEOREM 5. *Let X be a divisor on A. Let $y, z \in T_l(A)$ and $\hat{z} \in T_l(\hat{A})$. Then:*

(i) $(y, \varphi_X y) = 0.$

(ii) $(y, \varphi_X z) = -(z, \varphi_X y).$

(iii) $(y, \hat{z}) = -(\hat{z}, \kappa_A y)$ *where κ_A is the canonical homomorphism of A onto its double dual* (Chapter V, § 2).

Proof: Our statement (i) is but a translation in the group $T_l(A)$ of Proposition 7. From it, we conclude that

$$0 = (y + z, \varphi_X(y + z)) = (y, \varphi_X z) + (z, \varphi_X y)$$

thus proving (ii). As to (iii), let us take a non-degenerate divisor X on A, i.e., a divisor such that φ_X is surjective. It is clear that (i) and (ii) are valid for the extended Tate group, and we can find $z \in E_l(A)$ such that $\hat{z} = \varphi_X z$. According to Proposition 10 of Chapter V, § 2 we have

$$\varphi_X = {}^t\varphi_X \kappa_A.$$

Using the preceding theorem, we find

$$(\hat{z}, \kappa_A y) = (\varphi_X z, \kappa_A y) = (z, {}^t\varphi_X \kappa_A y) = (z, \varphi_X y).$$

Formula (iii) is then an immediate consequence of (ii).

Let us conclude this chapter with an application of the corollary of Theorem 4. Let X be a divisor on A, and denote by $C(X)$ the class of all divisors X' on A such that there exist two integers m, $m' > 0$ for which we have $mX \equiv m'X'$. We shall say that $C(X)$ determines a *polarization* of A if it contains a polar divisor, or equivalently if it contains an ample divisor. The class $C(X)$ is uniquely determined by one of its divisors. We then say that the couple $(A, C(X))$ determines a *polarized abelian variety*.

Let $\alpha : A \to B$ be a homomorphism, Y a divisor on B, and $X = \alpha^{-1}(Y)$ (which we may assume defined, after a generic translation). We know from Proposition 11 of Chapter V, § 2 that

$$\varphi_X = {}^t\alpha \varphi_Y \alpha.$$

If we take $A = B$ and if α is a birational automorphism of A, we shall say that α is an *automorphism of the polarized abelian variety* if there exist integers $m, m' > 0$ such that $mX \equiv m' \cdot \alpha^{-1}(X)$, or in other words if α^{-1} maps $C(X)$ into $C(\alpha^{-1}(X))$. This condition amounts to saying that

$$m \cdot \varphi_X = m' \cdot {}^t\alpha \varphi_X \alpha.$$

Taking the degree of both sides, and using the corollary of Theorem 4, we find $m = m'$. Hence we can rewrite our condition as $\alpha'\alpha = \delta_A$ where α' is the involution of Chapter V, § 2. We must therefore have $\mathrm{tr}(\alpha'\alpha) = \mathrm{tr}(\delta_A) = 2r$. Since $\mathrm{tr}(\alpha'\alpha)$ is a positive definite quadratic form and since the additive group End (A) is finitely generated, we have proved the following result.

PROPOSITION 8. *The group of birational automorphisms of a polarized abelian variety is finite.*

Historical note:

The development of the theory of l-adic representations follows that given by Weil [85], except that we have used a canonical and elegant construction of Tate, which allows us to work with l-adic spaces rather than with matrices.

The search of p-adic representations of dimension $2r$ is of considerable interest for a variety of reasons, some of them arising from complex multiplication, and others of a general nature. On one hand, Barsotti [7], [9] discovers certain representations in connection with the problems related to the existence of p-torsion and of the duality theory, and on the other hand, Serre [76] obtains analogues of the representations on the homology groups

$H^{p,o}$. For curves, he obtains a representation of complementary dimension to that given by the points whose order is a power of p on the Jacobian (i.e., the sum of the dimensions is equal to $2g$). The Witt vectors [97] play an important role in these matters.

The theorem concerning the finiteness of automorphisms of a polarized abelian variety is due to Matsusaka, and we have given here a proof due to Weil [96]. For a study of arbitrary polarized varieties, their automorphisms, and their fields of definition, see Matsusaka [61].

CHAPTER VIII

Algebraic Systems of Abelian Varieties

Chow has given a construction of the Picard and Albanese varieties as a consequence of the theory of algebraic systems of abelian varieties. Here, where we have given a direct construction, Chow's theory is rather an important complement to the general theory.

The notion of an algebraic system of abelian varieties is expressed algebraically by giving an abelian variety A_u defined over a regular extension $k(u)$ of a field k. The special members of the system are then the specializations of A_u over k. Actually, we are not concerned with these specializations here, but we merely wish to associate with our system its "fixed" part, that which does not depend effectively on the parameters. One does this from the point of view of the universal mapping property, for abelian varieties defined over k. Since we can take mappings into A_u or from A_u, we obtain in fact two fixed parts, called the trace and the image. They turn out to be isogenous, and there is a duality theory between them.

We also obtain a systematic analysis of the relation which exists between the Albanese variety of a variety V and that of its generic hyperplane section. If $\dim V \geqq 3$, we recover the classical theorem of Lefschetz according to which these two Albanese varieties are isomorphic. The dual statement for the Picard varieties is obtained from a general duality theory. For that purpose, we shall give a more precise formulation of the theorem concerning the transpose of an exact sequence than was given in Chapter V.

The first three sections depend only on Chapter II and could have been inserted immediately after our discussion of the Albanese variety, granting the fact that $\nu(n\delta) > 0$.

§ 1. *The K/k-image*

We recall that an extension K of a field k is said to be primary if the algebraic closure of k in K is purely inseparable. In characteristic 0, a primary extension is then regular. In general, there exists a purely inseparable extension k' of k such that Kk' is regular over k'. Thus a primary extension can be studied by taking up the two cases of regular extensions and purely inseparable extensions separately.

Our basic tool in this chapter is Chow's theorem (Theorem 5 of Chapter II, § 1). If A is defined over k, and if B is an abelian subvariety of A defined over a primary extension of k, then B is defined over k.

THEOREM 1. *Let K be a primary extension of k and let A be an abelian variety defined over K. Then there exists an abelian variety A_0, defined over k, and a surjective homomorphism $\lambda : A \to A_0$ defined over K satisfying the following condition. If B is an abelian variety defined over k, and if $\alpha : A \to B$ is a homomorphism defined over K, then there exists a homomorphism $\alpha_* : A_0 \to B$ defined over k such that the following diagram is commutative:*

$$\begin{array}{ccc} A & \xrightarrow{\lambda} & A_0 \\ & \alpha \searrow \quad \swarrow \alpha_* & \\ & B & \end{array}$$

Proof: The dimension of abelian varieties B such that there exists a homomorphism α of the above type is bounded. In addition, if $\alpha : A \to B$ and $\alpha' : A \to B'$ are defined over K, with B, B' defined over k, then the image of (α, α') in $B \times B'$ is an abelian subvariety of $B \times B'$ defined over k by Chow's theorem. We see therefore, that we can always form the supremum of two homomorphisms α, α' in an obvious sense. Thus we obtain a tower of rational homomorphisms. If we start our tower with an abelian variety B of maximal dimension, it is clear that there cannot be an infinite tower above it, and we see that a map $\lambda : A \to A_0$

which is maximal from the birational point of view satisfies the desired condition.

We shall say that (A_0, λ) in Theorem 1 is a *K/k-image* of A. It is clear that two such images differ only by a birational isomorphism, defined over k.

PROPOSITION 1. *Let $k \subset K_1 \subset K_2$ be a tower such that K_1 is primary over k and K_2 primary over K_1. Let A_2 be an abelian variety defined over K_2. Let (A_1, λ_1) be the K_2/K_1-image of A_2, and (A, λ) the K_1/k-image of A_1. Then $(A, \lambda\lambda_1)$ is the K_2/k-image of A_2.*

Proof: Obvious from the definitions.

As a special case, we get:

PROPOSITION 2. *Let K be a primary extension of k, and let A be defined over K. Let K' be a primary extension of K. Then a K/k-image (A_0, λ) of A is also a K'/k-image of A.*

PROPOSITION 3. *Let K be a primary extension of k, and let A be an abelian variety defined over K. Let (A_0, λ) be its K/k-image. Then λ is primary, or in other words, the kernel of λ is connected.*

Proof: We shall reduce our proposition to the case where λ is an isogeny. Let $\varphi : A \to C$ be the canonical homomorphism of A onto the factor group by the connected component B of the kernel of λ. By Chow's theorem, B is defined over K, and by definition, φ is separable. We have a homomorphism $\lambda' : C \to A_0$ defined over K such that the following diagram is commutative.

$$\begin{array}{ccc} & A & \\ \varphi \swarrow & & \searrow \lambda \\ C & \xrightarrow{\lambda'} & A_0 \end{array}$$

The canonical map φ is regular, and it will suffice to show that λ' is purely inseparable. I contend that (A_0, λ') is the K/k-image of C. Indeed, if $\beta : C \to C_0$ is the canonical map of C onto its K/k-image, the map $\beta\varphi$ gives a homomorphism of A onto an abelian variety defined over k. This map can be factorized through λ, that is to say there exists a homomorphism $(\beta\varphi)_* : A_0 \to C_0$ such that $\beta\varphi = (\beta\varphi)_*\lambda$.

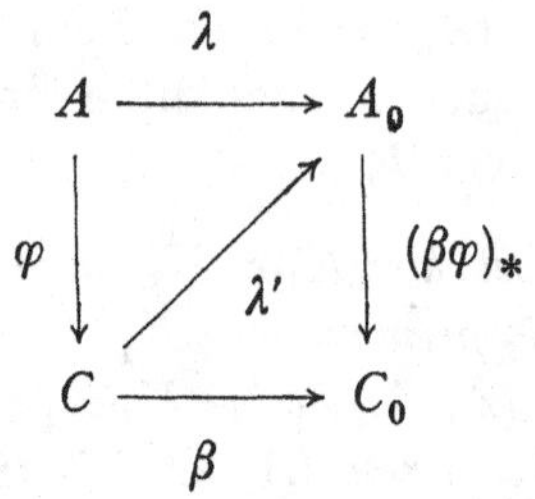

There exists a homomorphism $h : C_0 \to A_0$ such that $\lambda' = h\beta$. Since by definition (A_0, λ) satisfies the universal mapping property, it follows that h must be a birational isomorphism. This means that C_0 can be identified with A_0, and thus brings us to considering the case where the kernel of λ is finite.

Assume therefore that in our proposition, the kernel of λ is finite. It will suffice to show that there exists a purely inseparable isomorphism $\lambda' : A \to A'$ defined over K, with A' defined over k. The universal mapping property of (A_0, λ) will then show that λ must be purely inseparable:

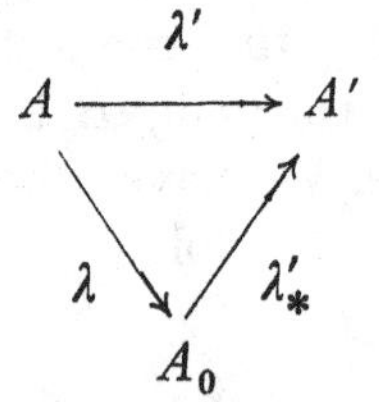

There exists an isogeny $\tau : A_0 \to A$ defined over K. Its kernel is finite and K-closed, hence k-closed. The factor group A_1 of A_0 by this kernel is defined over a purely inseparable extension k_1 of k, and there exists an isomorphism of this factor group onto A. Raising A_1 to a suitable power p^m of the characteristic, we find an isomorphism (always purely inseparable, of course) of A onto $A_1^{(p^m)}$, defined over K. This is what we wanted.

Using Serre's Galois' theory of purely inseparable p-extensions of abelian varieties, one sees immediately that if K is purely inseparable over k, or regular over k, then the canonical map λ onto the K/k-image is not necessarily birational.

One can prove for the K/k-image results which are analogous to those concerning the field of definition of the Albanese variety, using the criteria of Chapter I, § 3. The proofs follow the same pattern as before.

To make the notation functorial, as we did with the Albanese variety, we can denote A by A_K when it is defined over K. We denote by $\mathrm{Im}_k^K(A_K)$ the set of all K/k-images (A_0, λ) of A_K. If F is an extension of k which is independent of K, then A_K is also defined over KF, and KF is primary over F.

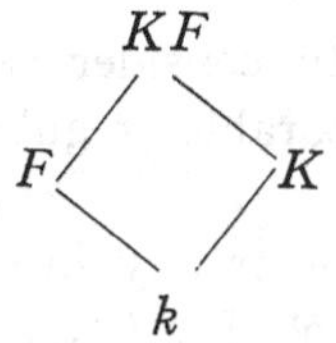

By an abuse of notation, we write

$$I_k^F : \mathrm{Im}_F^{KF}(A_K) \to \mathrm{Im}_k^K(A_K)$$

the canonical homomorphism derived from the universal mapping property of the KF/F-image of $A_K = A_{KF}$. We shall prove below that I_k^F is always an isomorphism (but may be purely inseparable).

PROPOSITION 4. *Let A_K be an abelian variety defined over a primary extension K of k. Let F be an extension of k which is independent of K over k, and let (A_F, λ) be a KF/F-image of A_K. Let E be a subfield of F containing k such that A_F is defined over E, and λ is defined over KE. Then (A_F, λ) is also a KE/E-image of A.*

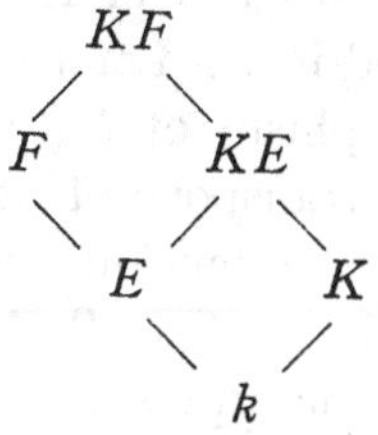

Proof: Let (B, β) be a KE/E-image of A_K. By definition, there exists a homomorphism $\alpha : B \rightarrow A_F$ defined over E such that $\lambda = \alpha\beta$, and there exists a homomorphism $\alpha' : A_F \rightarrow B$ defined over F such that $\beta = \alpha'\lambda$. We get $\lambda = \alpha\alpha'\lambda$ and $\beta = \alpha'\alpha\beta$, which proves that $\alpha\alpha'$ and $\alpha'\alpha$ are equal to the identity, and hence that α, α' are birational.

THEOREM 2. *Let K be a primary extension of k, and let A_K be defined over K. Let F be an extension of k, independent of K over k. Then the canonical homomorphism I_k^F is always a purely inseparable isomorphism. If F is separable over k, then it is a birational isomorphism.*

Proof: We may clearly consider separately the three cases where F is purely inseparable, regular, or separable algebraic over k.

Suppose first that F is purely inseparable over k. We may assume it of finite degree p^m. Let (A_1, λ_1) be a KF/F-image of A_K. Then $A_1^{(p^m)}$ is defined over k, because $k_1^{p^m}$ is contained in k. Since the homomorphism of A_1 onto $A_1^{(p^m)}$ is purely inseparable, one sees immediately that I_k^F is purely inseparable.

We turn to the case of regular extensions. For the rest of the proof, we let F be a universal domain for separable extensions of k, i.e., a separable extension of k of infinite transcendence degree and separably closed. In view of Proposition 4 it will suffice to prove that there exists a KF/F-image of A_K which is defined over k. There is a KF/F-image of A_K which is defined over a finitely generated extension of k which we assume to be regular. We denote it by $k(v)$, and let (A_v, λ_v) be the $K(v)/k(v)$-image of A_K. By Proposition 4 it is also a KF/F-image. Let v' be a generic specialization of v over K, independent of v, and such that $k(v')$ is contained in F. Let $(A_{v'}, \lambda_{v'})$ be the transform of (A_v, λ_v) by the isomorphism of $K(v)$ which leaves K fixed and sends v into v'. By transport of structure, $(A_{v'}, \lambda_{v'})$ is a $K(v')/k(v')$-image of A_K. The extension $k(v, v')$ of k is regular, and (A_v, λ_v) is *a fortiori* defined over $k(v, v')$. According to Proposition 4, there exist birational isomorphisms

$$f_{v,v'} : A_{v'} \to A_v,$$
$$f_{v',v} : A_v \to A_{v'}$$

defined over $k(v, v')$ satisfying the commutative diagram:

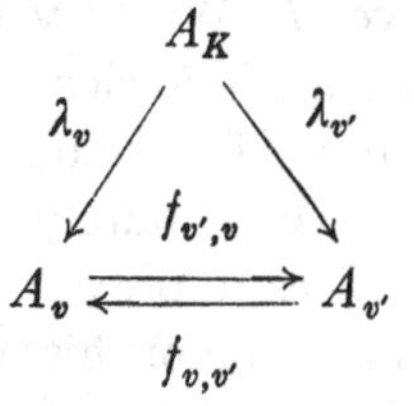

and one sees immediately that the coherence conditions of Chapter I, § 3 are satisfied, by means of the uniqueness of the $f_{v,v'}$ and $f_{v',v}$. Hence there exists an abelian variety A_0 defined over k which becomes birationally isomorphic to A_v over $k(v)$. It is clear from Theorem 1G of Chapter I, § 3 that we can find a homomorphism $\lambda : A_K \to A_0$ defined over K such that (A_0, λ) is a $K(v)/k(v)$-image of A_K, and hence a K/k-image of A_K.

Let us finally consider the case where the KF/F-image of A_K is defined over a finite separable extension k_1 of k. Denote by $\sigma, \tau, \ldots$ the automorphisms of the separable closure of k leaving k fixed. Each σ can be extended to an automorphism of the universal domain leaving K fixed. Denote by (A_1, λ_1) the KF/F-image of A_K such that A_1 is defined over k_1 and λ_1 over Kk_1, and denote by $(A_1{}^\sigma, \lambda_1{}^\sigma)$ its transform by σ. The arguments used above can be applied to the present case *mutatis mutandis*, replacing (v, v') by (σ, τ) and (A_v, λ_v) by $(A_1{}^\sigma, \lambda_1{}^\sigma)$. We obtain birational isomorphisms $f_{\sigma,\tau}$ satisfying the coherence conditions of Theorem 2, Chapter I, § 3 and thus we obtain again an abelian variety A_0 defined over k and a homomorphism $\lambda : A_K \to A_0$ defined over K such that (A_0, λ) is a Kk_1/k_1-image of A. This proves our theorem.

COROLLARY 1. *Let K be a primary extension of k, and let A_K be defined over K. Let E be an extension of k independent of K. Suppose that there exists an abelian variety B defined over E, and a surjective homomorphism $\beta : B \to A_K$ defined over KE. Let (A_0, λ) be the K/k-image of A_K. Then λ is an isomorphism.*

Proof: By Theorem 2 it suffices to prove our corollary in the case where $E = k$. Let C be the connected component of the kernel of β. Then C is defined over a purely inseparable extension of K, and hence over K by Chow's theorem. Once again by this theorem, it is defined over k. We may therefore assume that β is an isogeny. Its finite kernel is K-closed, and since K is primary over k, it is therefore k-closed. The factor group of B by the kernel of β is therefore defined over a purely inseparable extension k_1 of k. By Theorem 2, it suffices to prove our theorem over the field k_1. In this case, there is an isomorphism of the preceding factor group onto A_K, defined over Kk_1. Hence there exists an isomorphism of A_K onto an abelian variety defined over k_1, and we see that our λ must be purely inseparable.

COROLLARY 2. *Let K be a primary extension of k, and E a separable extension of k, independent from K over k. Let $A = A_K$ be an abelian variety defined over K, and $B = B_E$ an abelian variety defined over E. Suppose that there exists a birational isomorphism of B onto A, defined over KE. Let (A_0, λ) be the K/k-image of A. Then λ is a birational isomorphism.*

Proof: By Theorem 2, the canonical isomorphism I_E^k is birational. We may therefore assume that $E = k$. Let $\alpha : A \to B$ be the birational isomorphism which is the inverse of the one given in the statement of the corollary. It is then clear that (B, α) is a K/k-image of A. This proves the corollary.

We now come to a theorem which is more refined than the preceding one, and allows us in some applications to use a birational isomorphism rather than a purely inseparable one. We call the next theorem the *regularity theorem.* (Cf. § 5 following.)

THEOREM 3. *Let A_u be an abelian variety defined over a finitely generated regular extension $k(u)$ of k, and let (B, λ_u) be a $k(u)/k$-image. Let $u_1, \ldots, u_m$ be independent generic specializations of u over k, and let (A_{u_i}, λ_{u_i}) be the transform of (A_u, λ_u) by the isomorphism $k(u) \to k(u_i)$. Let $A_{(m)}$ be the product $A_{u_1} \times \ldots \times A_{u_m}$. Then for all sufficiently large m, the homomorphism $F_{(m)} : A_{(m)} \to B$ such that*

$$F_{(m)}(x_1, \ldots, x_m) = \sum_{i=1}^{m} \lambda_{u_i}(x_i)$$

is regular.

Proof: Let

$$\lambda_{(m)} : A_{(m)} \to B_{(m)}$$

be the product of the λ_{u_i}. Let s_m be the sum, from $B_{(m)}$ to B. By definition, we have $F_{(m)} = s_m \lambda_{(m)}$. According to Proposition 3 each λ_{u_i} is primary, that is to say, its kernel is connected. Furthermore, s_m is regular. Indeed, if $y_1, \ldots, y_m$ are independent generic points of B, and $y = y_1 + \ldots + y_m = s_m(y_1, \ldots, y_m)$ then $k(y, y_2, \ldots, y_m) = k(y_1, \ldots, y_m)$ is a regular extension of $k(y)$. Its kernel is therefore also connected, and hence so is the kernel of $F_{(m)}$, which we denote by $X_{(m)}$. Let

$$\varphi_{(m)} : A_{(m)} \to C_{(m)}$$

be the canonical map of $A_{(m)}$ onto the factor group $C_{(m)}$ of $A_{(m)}$ by $X_{(m)}$. Since $X_{(m)}$ is an abelian subvariety of $A_{(m)}$ and is defined over a purely inseparable extension of

$$k_{(m)} = k(u_1, \ldots, u_m)$$

it is defined over $k_{(m)}$ by Chow's theorem. The map $\varphi_{(m)}$ is therefore defined over $k_{(m)}$ and is also regular. We can factorize $F_{(m)}$ and write

$$F_{(m)} = \tau_{(m)}\varphi_{(m)}$$

where $\tau_{(m)}$ is purely inseparable, and defined over $k_{(m)}$. Our theorem will be proved if we show that for all sufficiently large m, the map $\tau_{(m)}$ is in fact regular, because a composition of surjective regular maps is regular:

$$\begin{array}{ccc}
A_{(m)} & \xrightarrow{\lambda_{(m)}} & B_{(m)} \\
\downarrow{\scriptstyle \varphi_{(m)}} & & \downarrow{\scriptstyle s_m} \\
C_{(m)} & \xrightarrow[\tau_{(m)}]{} & B
\end{array}$$

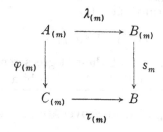

The main step of our proof will consist in showing that $C_{(m)}$ is birationally equivalent, over $k(u_1, \ldots, u_m)$, to an abelian variety defined over k. This will be done by a symmetry argument, and Corollary 2 of Theorem 2.

Let $u_{m+1}, \ldots, u_n$ be independent generic specializations of u over $k(u_1, \ldots, u_m)$ and write

$$A'_{(n)} = A_{u_{m+1}} \times \ldots \times A_{u_{m+n}}.$$

With an analogous notation, we define $\lambda'_{(n)}$ and $C'_{(n)}$ as above, but relative to the indices $m+1, \ldots, m+n$.

We can embed $A_{(m)}$ into $A_{(m)} \times A'_{(n)}$ by $(\delta, 0)$, where δ is the identity. The following diagram is then commutative:

$$\begin{array}{ccccc}
A_{(m)} & \xrightarrow{(\delta,\,0)} & A_{(m)} \times A'_{(n)} & \xrightarrow{\lambda_{(m)} \times \lambda_{(n)}} & B_{(m)} \times B_{(n)} \\
\downarrow{\scriptstyle \varphi_{(m)}} & & \downarrow & & \downarrow{\scriptstyle (s_m,\, s_n)} \\
C_m & \xrightarrow[(\delta,\,0)]{} & C_{(m)} \times C'_{(n)} & \xrightarrow{\tau_{(m)} \times \tau'_{(n)}} & B \times B \\
 & & \downarrow{\scriptstyle h} & & \downarrow{\scriptstyle s_2} \\
 & & C_{(m+n)} & \xrightarrow[\tau_{(m+n)}]{} & B
\end{array}$$

In this diagram, h denotes the canonical homomorphism obtained from the factor group. We see immediately that the composed map

$$\tau_{(m+n)}\, h(\delta,\ 0)$$

is equal to $\tau_{(m)}$ on $C_{(m)}$. This shows that the degrees of the maps τ decrease monotonically, i.e., that

$$\nu(\tau_{(m)}) \geqq \nu(\tau_{(m+n)}).$$

Let us take m large, so that these degrees remain constant for all n. In this case, we see that $h(\delta, 0)$ is a birational isomorphism on $C_{(m)}$.

Since $C_{(m)}$ and $C_{(m+n)}$ are defined over $k(u_1, \ldots, u_m, u_{m+1}, \ldots, u_{m+n})$

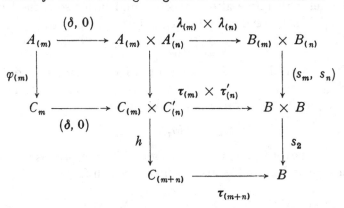

and since the graph of $h(\delta, 0)$ is an abelian subvariety of their product, it follows from Chow's theorem that $h(\delta, 0)$ is defined over that field.

Take $m = n$. Then $C_{(m)}$ is transformed into $C'_{(m)}$ by the isomorphism $k(u_1, \ldots, u_m) \to k(u_{m+1}, \ldots, u_{2m})$ and these two groups are birationally isomorphic over the field composed of these two fields, since by the preceding remark, they are both birationally isomorphic to $C_{(m+m)}$ over the field $k(u_1, \ldots, u_m, u_{m+1}, \ldots, u_{m+m})$. According to Corollary 2 of Theorem 2 there exists an abelian variety C defined over k and a birational isomorphism

$$g_{(m)} : C_{(m)} \to C$$

defined over $k_{(m)}$ for all sufficiently large m.

Let us apply these results replacing m by n. In our big diagram, take $m = 1$ and take n large. Then

$$\alpha = g_{(n+1)}\, h(\delta, 0)$$

is a homomorphism of $C_{(1)}$ onto C. This homomorphism is defined over $k_{(n+1)}$ which is a regular extension of k_1. Since its graph is an abelian subvariety of the product, it is defined over $k_{(1)}$ by Chow's theorem. Put $\beta = \tau_{(1+n)} g_{(1+n)}{}^{-1}$. It is a homomorphism of C onto B, defined over $k_{(1+n)}$ and hence over k. The following diagram is commutative:

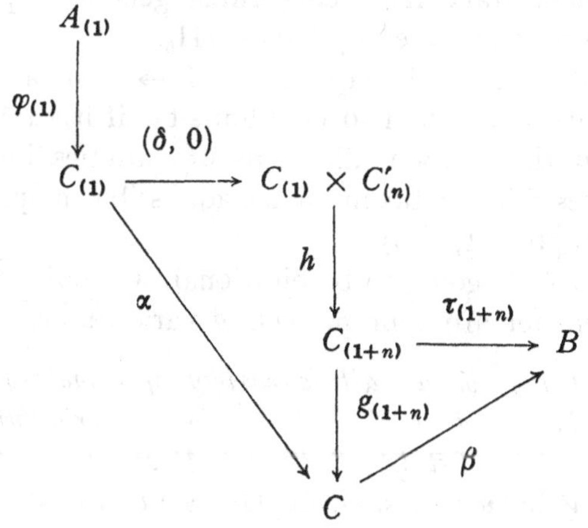

In view of the universal mapping property of B, and of the fact that

$$\tau_{(1+n)}h(\delta, 0) = \tau_{(1)}$$

and hence that $\varphi_{(1)}$ followed by this homomorphism is equal to λ, it follows that $\alpha\varphi_{(1)}$ can dominate B by β only if β is birational, and in this case $\tau_{(1+n)}$ is birational. This proves the regularity theorem.

COROLLARY. *Let K be a regular extension of k, and A an abelian variety defined over K. Let E be any extension of k, independent from K over k. Let (A_0, λ) be the K/k-image of A. Then it is also the KE/E-image of A.*

Proof: We may clearly assume that $K = k(u)$ is finitely generated. Let (A_0', λ') be the KE/E-image of A. Let $F_{(m)}' : A_{(m)} \to A_0'$ and $F_{(m)} : A_{(m)} \to A_0$ be the homomorphisms as in the theorem, relative to E and k respectively. Then $F_{(m)}'$ and $F_{(m)}$ are regular for large m. Let

$$I_k^E : A_0' \to A_0$$

be the canonical map. Then we obviously have $F_{(m)} = I_k^E F_{(m)}'$. From this we can conclude that I_k^E is also regular.

§ 2. *The generic hyperplane section*

For all the elementary facts concerning generic hyperplane sections of a variety we refer to IAG—VII_6.

We recall that a rational map $f : V \times V \to A$ of a product into an abelian variety is said to be admissible if it vanishes on the diagonal. In this section, all maps of varieties into their Albanese varieties will be taken to be admissible maps of the product (see Chapter II, § 3).

Since our results are going to be birational, we could state the following theorem for affine or projective varieties, as we wish.

THEOREM 4. *Let V be an affine variety of dimension $\geqq 2$, defined over a field k, and let $W_u = V \cdot L_u$ be a generic hyperplane section defined over k_u. Let (A, f) be a k-Albanese variety of V, and let $i : W_u \to V$ be the inclusion. Let (B_u, g_u) be a $k(u)$-Albanese* *variety of* W_u, *and let* $i_* : B_u \to A$ *be the induced homomorphism which makes the following diagram commutative*:

$$\begin{array}{ccc} W_u \times W_u & \xrightarrow{\ i\ } & V \times V \\ {\scriptstyle g_u}\downarrow & & \downarrow{\scriptstyle f} \\ B_u & \xrightarrow[\ i_*\ ]{} & A \end{array}$$

Then (A, i_*) *is a* $k(u)/k$*-image of* B_u.

Proof: Let (B, λ_u) be a $k(u)/k$-image of B_u. Let (x, y) be a generic point of $W_u \times W_u$ over $k(u)$, and hence a generic point of $V \times V$ over k (an elementary fact about generic sections). The point $\lambda_u g_u(x, y)$ is rational over $k(x, y, u)$ which is a purely transcendental extension of $k(x, y)$. It is a point of the abelian variety B, which we may interpret as giving a rational map of a pure variety into B, defined over $k(x, y)$. By the corollary of Theorem 4, Chapter II, § 1 we conclude that this map is constant. This means that $\lambda_u g_u(x, y)$ is rational over $k(x, y)$. Hence there exists a rational map $h : V \times V \to B$ defined over k, such that

$$h(x, y) = \lambda_u g_u(x, y).$$

Furthermore, h is admissible since $h(x, x) = \lambda_u g_u(x, x) = 0$.

The homomorphism i_* being defined over $k(u)$, we get from the definition of the $k(u)/k$-image a homomorphism

$$\beta : B \to A$$

such that $i_* = \beta\lambda_u$, or in other words, $fi = \beta\lambda_u g_u$. By the definition of h, we have $hi = \lambda_u g_u$. Hence we find $fi = \beta hi$, or also

$$fi(x, y) = \beta hi(x, y).$$

Therefore $f = \beta h$. Since f satisfies the universal mapping property for admissible maps of $V \times V$ into abelian varieties, it follows that β is a birational isomorphism, as was to be shown.

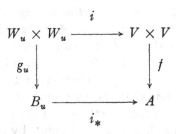

As already pointed out, the above theorem applies to projective varieties as well, and this allows us to give an important complement to the theorems concerning the field of definition of the Albanese variety.

COROLLARY 1. *Let V be a projective variety, defined over k, and non-singular in codimension 1. Let (A, f) be a k-Albanese variety of V. Then (A, f) is also an E-Albanese variety of V for every field E containing k, i.e., it is an Albanese variety of V.*

Proof: We can proceed by induction on the dimension of V. For a curve, we know that the theorem is true because the Jacobian can be constructed directly over a separable extension of k over which the curve has a rational point, by means of the symmetric product of the function field of the curve. Let now V be of arbitrary dimension $\geqq 2$, and let L_u be a generic hyperplane section over k. Then one knows that $V \cdot L_u$ is also non-singular in codimension 1, and hence that its $k(u)$-Albanese variety is an Albanese variety. We can choose u generic over E. The preceding theorem, combined with the corollary of Theorem 3, § 1 gives us the desired result.

COROLLARY 2. *Let A be an abelian variety. Then there exists a product of Jacobians $\prod J_i$ and a surjective homomorphism of this product onto A which is a regular map.*

Proof: This is a consequence of Theorems 3 and 4, and of the transitivity of the K/k-image (Proposition 1 of § 1).

THEOREM 5. *Let V be an affine variety defined over k, and let L_u be a generic hyperplane over k. Let $W_u = V \cdot L_u$ and let $\lambda_u : B_u \to A$ be the canonical homomorphism of the $k(u)$-Albanese variety of W_u onto the k-Albanese variety of V. If* $\dim V \geqq 3$, *then λ_u is an isomorphism.*

Proof: Let $L_{u'}$ be a generic hyperplane such that u and u' are independent over k. Then $L_u \cdot L_{u'}$ is defined over an extension $k(w)$ contained in $k(u, u')$, and such that $k(u, u')$ is regular over $k(w)$. This is easily seen, either by taking the Chow coordinates (Plücker coordinates in this case) of $L_u \cdot L_{u'}$, or by making a direct analysis of the smallest field of definition of $L_u \cdot L_{u'}$. Let $A_{u,u'}$ be the $k(u, u')$-Albanese variety of $V \cdot L_u \cdot L_{u'}$. Since $V \cdot L_u \cdot L_{u'}$ is defined over $k(w)$, we can choose

$A_{u,u'}$ also defined over $k(w)$ by Theorem 12 of Chapter I, § 3. We know that B_u is the $k(u, u')/k(u)$-image of $A_{u,u'}$ by Theorem 4. Let $\varphi : A_{u,u'} \to B_u$ be the canonical homomorphism defined over $k(u, u')$. Let σ be the automorphism of $k(u, u')$ which permutes u and u'. Then σ leaves $L_u \cdot L_{u'}$ fixed, and hence leaves $k(w)$ fixed. Moreover, by symmetry, the transform of φ by σ gives us the canonical homomorphism

$$\varphi^\sigma = \varphi' : A_{u,u'} \to B_{u'}$$

of $A_{u,u'}$ onto its $k(u, u')/k(u')$-image $B_{u'}$, which is the $k(u')$-Albanese variety of $V \cdot L_{u'}$.

Now let C be the connected component of the kernel of φ. Then C is defined over a purely inseparable extension of $k(u, u')$ and hence over $k(w)$ by Chow's theorem. It follows that C is left invariant by σ. This implies that C is also the connected component of the kernel of φ^σ, and we thus obtain an isogeny of the factor group A/C onto B_u and $B_{u'}$, defined over $k(u, u')$. We see that B_u and $B_{u'}$ are isogenous over $k(u, u')$. According to Corollary 1 of Theorem 2, § 1 it follows that λ_u is an isomorphism.

It is not known, at the time this book is written, whether λ_u may sometimes be purely inseparable.

§ 3. *The K/k-trace*

We study in this section the dual construction of the K/k-image, that is to say, the existence of a fixed abelian variety satisfying the universal mapping property for mappings of abelian varieties into a given one, depending on parameters.

THEOREM 6. *Let K be a primary extension of k, and A an abelian variety defined over K. Then there exists a smallest abelian subvariety A^* of A, defined over K satisfying the following property: If B is an abelian variety defined over an extension E of k, independent of K over k, and if $\beta : B \to A$ is a homomorphism of B into A, defined over KE, then the set-theoretic image $\beta(B)$ of B is contained in A^*.*

Proof: We are going to see that A^* is simply the sum of all the abelian subvarieties of A which are of type $\beta(B)$. To begin with, note that in the statement of our theorem, we can take E free from any given extension, after having taken a suitable isomorphism of it. Consequently, if we are given two homomorphisms $\beta_1 : B_1 \to A$ and $\beta_2 : B_2 \to A$ defined over E_1K and E_2K, respectively, such that E_1 and E_2 are each free from K over k, then we may assume in addition, without loss of generality, that they are free from each other, and that their compositum E_1E_2 is free from K. We then obtain a homomorphism $\beta_3 : B_1 \times B_2 \to A$ defined over $E_3 = E_1E_2$ mapping $B_1 \times B_2$ onto the sum $\beta_1(B_1) + \beta_2(B_2)$, and β_3 is defined over E_3K. Since the dimension of the abelian subvarieties of A is bounded by that of A, we get immediately the existence of our variety A^*.

In addition, we also see that there exists a surjective homomorphism $\beta : B \to A^*$, with B defined over some extension E of k, free from K over k, and β defined over KE. By the theory of the K/k-image (Corollary 1, of Theorem 2, § 1) it follows that the canonical homomorphism of A^* onto its K/k-image is an isomorphism.

We are going to see that A^* is isogenous to the K/k-image (A_0, λ) of A. By the Poincaré complete reducibility theorem, there exists an isogeny defined over K which maps A_0 into A, and hence into A^*. This shows that $\dim A^* \geqq \dim A_0$. On the other hand, since we have seen that the canonical homomorphism of A^* onto its K/k-image is an isomorphism, the same theorem shows that we can find a homomorphism of A, defined over K, onto an abelian variety defined over k, of dimension $\geqq \dim A^*$. Consequently, $\dim A_0 \geqq \dim A^*$, and we have the equality. It follows that A_0 and A^* are isogenous over K.

Let us call A^* the *K/k-maximal abelian subvariety* of A. We have proved the following result.

THEOREM 7. *Let K be a primary extension of k. Let A be an abelian variety defined over K, and A^* its K/k-maximal abelian sub-variety. Then the canonical homomorphism of A^* onto its K/k-image is an isomorphism, and A^* is isogenous over K to the K/k-image of A.*

The next proposition is obvious.

PROPOSITION 5. *Denote by $S_{K/k}(A)$ the variety A^* of* Theorem 6. *Let E be an extension of k, free from K over k. Then $S_{KE/E}(A) =$*

$S_{K/k}(A)$. *If* $k \subset K \subset K_1$ *is a tower of primary extensions, then* $S_{K/k}(A) = S_{K_1/k}(A)$.

We now come to the trace. Let K be a primary extension of k, and let A be an abelian variety defined over K. We shall say that a couple (A', τ) consisting of an abelian variety A' defined over k, and a homomorphism $\tau : A' \to A$ defined over K is a *K/k-trace* of A if τ has finite kernel and satisfies the following condition. Given an abelian variety B defined over k and a homomorphism $\beta : B \to A$ defined over K, there exists a homomorphism $\beta' : B \to A'$ defined over k such that the following diagram is commutative.

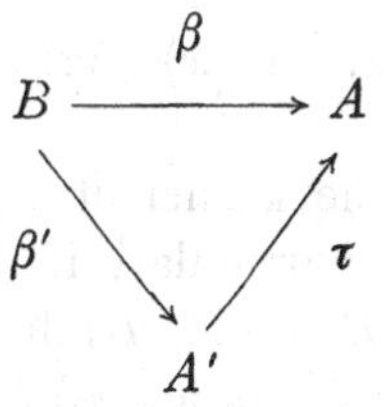

Using the fact that τ has finite kernel, one sees immediately that β' is uniquely determined by the above condition.

THEOREM 8. *Let K be a primary extension of k, and let A be an abelian variety defined over K. Then there exists a K/k-trace of A.*

Proof: The construction is entirely similar to that of the K/k-image, taking the sup of homomorphisms of abelian varieties into A, and using Chow's theorem. We leave the details to the reader.

Contrary to what happens to the image, it is not necessarily true that τ is purely inseparable. It is easy to construct examples, using elliptic curves with points of order p.

PROPOSITION 6. *Let $k \subset K_1 \subset K_2$ be a tower such that K_1 is primary over k and K_2 is primary over K_1. Let A_2 be an abelian variety defined over K_2. Let (A_1, τ_1) be a K_2/K_1-trace of A_2, and let (A, τ) be a K_1/k-trace of A_1. Then $(A, \tau_1\tau)$ is a K_2/k-trace of A_2.*

Proof: Obvious from the definitions and Chow's theorem.

PROPOSITION 7. *Let K be a primary extension of k, and let A be defined over K. Let K' be a primary extension of K. Then a K/k-trace of A is also a K'/k-trace of A.*

Proof: Obvious from Chow's theorem.

Finally, we come to the regularity theorem.

THEOREM 9. *Let $k(u)$ be a finitely generated regular extension of k, and let A_u be an abelian variety defined over $k(u)$. Let B be an abelian variety defined over k, and $\tau_u : B \to A_u$ a homomorphism with finite kernel, defined over $k(u)$. Let $u_1, \ldots, u_m$ be generic independent specializations of u over k, and A_{u_i}, τ_{u_i} the transforms of A_u, τ_u under the map $u \to u_i$. Then (B, τ_u) is a $k(u)/k$-trace of A_u if and only if τ_u maps B onto A_u^*, and for m large, the homomorphism*

$$T_m : B \to A_{u_1} \times \ldots \times A_{u_m}$$

such that $T_m(x) = (\tau_{u_1}x, \ldots, \tau_{u_m}x)$ gives a birational correspondence between B and $T_m(B)$.

Proof: Observe that the kernel of T_m is the same as that of τ_u and of each τ_{u_i}. In particular, if T_m becomes a birational correspondence between B and $T_m(B)$ then τ_u is purely inseparable.

Assume first that (B, τ_u) is a $k(u)/k$-trace of A_u. For $m \geqq n$ we have a natural projection of $T_m(B)$ onto $T_n(B)$:

$$T_m(B) \to T_n(B) \to T_1(B) = \tau_{u_1}(B),$$

and its degree must eventually become equal to 1 because it is bounded by the degree of B over $T_n(B)$. Take $m = 2n$ large. Then $T_{2n}(B)$ and $T_n(B)$ are birationally equivalent over $k(u_1, \ldots, u_{2n})$, so that the transform of $T_n(B)$ under the isomorphism of $k(u_1, \ldots, u_n)$ which maps $(u_1, \ldots, u_n)$ on $(u_{n+1}, \ldots, u_{2n})$ is birationally isomorphic to $T_n(B)$ over $k(u_1, \ldots, u_{2n})$. By Corollary 2 of Theorem 2, § 1 it follows that there exists an abelian variety B' defined over k and a birational isomorphism $\beta_n : B' \to T_n(B)$ defined over $k(u_1, \ldots, u_n)$. The isomorphism $\beta = \beta_n^{-1} T_n : B \to B'$ is defined over k by Chow's theorem, and we have commutativity in the following diagram.

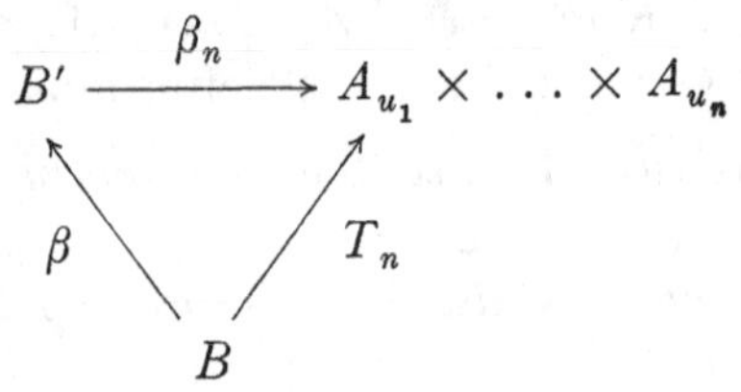

Taking the projection on the first factor, we get

$$(pr_1 \beta_n)\beta = \tau_{u_1},$$

and thus a commutative diagram

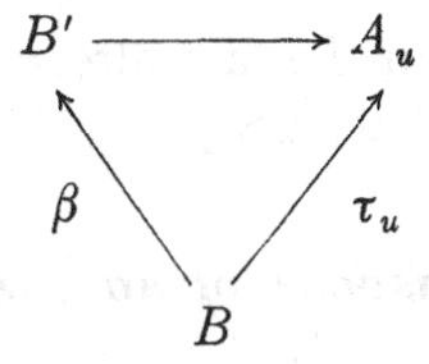

Since we assumed (B, τ_u) to be a $k(u)/k$-trace, it follows that β is a birational isomorphism. From this it is clear that (B, τ_u) satisfies the conditions of our theorem.

Conversely, assume that (B, τ_u) satisfies these conditions. Let (A', φ_u) be a $k(u)/k$-trace of A. Let

$$\Phi_m : A' \to A_{u_1} \times \ldots \times A_{u_m}$$

be given by the formula $\Phi_m(y) = (\varphi_{u_1} y, \ldots, \varphi_{u_m} y)$. Since by hypothesis we have $\tau_u(B) = A_u^* = \varphi_u(A')$, it follows that $T_m(B) = \Phi_m(A')$. The homomorphism $T_m^{-1}\Phi_m : A' \to B$ is defined over k by Chow's theorem, and we have a commutative diagram

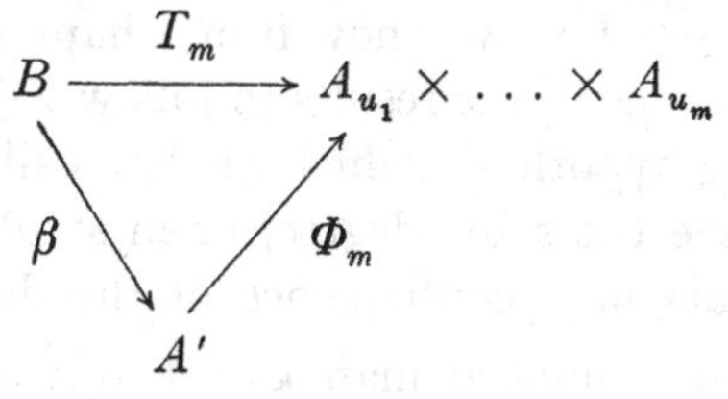

Taking the projection on the first factor and using the fact that (A', φ_u) is a $k(u)/k$-trace, we see that β must be birational. This proves what we wanted.

The condition of Theorem 9 is independent of the ground field. This yields

COROLLARY 1. *Let K be a regular extension of k, A an abelian variety defined over K, and (B, τ) a K/k-trace of A. Let E be an* *extension of k, independent of K. Then (B, τ) is also a KE/E-trace of A.*

Proof: The reduction to the finitely generated case is immediate.

COROLLARY 2. *Let the notation be as in Corollary 1. Then τ is purely inseparable.*

Proof: This comes from the fact already noted that the kernel of T_m is the same as that of τ_u.

§ 4. The transpose of an exact sequence

We are going to complete our results on exact sequences. If we are given a sequence of homomorphisms of abelian varieties

$$0 \to A \xrightarrow{\alpha} B \xrightarrow{\beta} C \to 0$$

we shall say that it is *exact* if it is exact in the set-theoretic sense. This implies in particular that α is purely inseparable and that the kernel of β is connected, equal to the image of α. We shall say that it is *regular* if α and β are regular, as rational maps. In this case, α establishes a birational correspondence between A and its image $\alpha(A)$, so that we may assume that α is an inclusion. Furthermore, β can be identified with the canonical homomorphism on the factor group of B by A.

If α is an isogeny, we know from Chapter V, § 1 that ${}^t\alpha$ is also an isogeny. Some of the results to follow will be established under the following hypothesis, which we shall call the *duality hypothesis*: α and ${}^t\alpha$ have the same degree, or in symbols, $\nu(\alpha) = \nu({}^t\alpha)$.

As an immediate consequence of the duality hypothesis, one sees that the canonical map $\kappa_A : A \to \hat{\hat{A}}$ is birational, by Proposition 5 of Chapter V. (See p. 229.)

THEOREM 10. *Let*

$$0 \to A \xrightarrow{\alpha} B \xrightarrow{\beta} C \to 0$$

be a regular exact sequence. Under the duality hypothesis, the transposed sequence

$$0 \leftarrow \hat{A} \xleftarrow{{}^t\alpha} \hat{B} \xleftarrow{{}^t\beta} \hat{C} \leftarrow 0$$

is exact and regular. If one assumes only that the torsion group on C is trivial, then the transposed sequence is exact, and ${}^t\beta$ *is regular.*

Proof: Let us first assume the duality hypothesis. Say we wish to prove that ${}^t\beta$ is regular. If it is not, we know from the rough results of Chapter V, § 1 that its kernel is finite. Hence we have a sequence of maps

$$\hat{C} \xrightarrow{\tau} {}^t\beta(\hat{C}) \xrightarrow{\sigma} \hat{B}$$

where $\sigma\tau = {}^t\beta$, where $\nu(\tau) > 1$ and where σ is an inclusion. Taking the transpose, we get $t^2\beta = {}^t\tau\,{}^t\sigma$. From Proposition 8 of Chapter V, § 2 and the duality hypothesis according to which κ_A is birational, we conclude that $t^2\beta$ is regular. From the duality hypothesis we have $\nu({}^t\tau) > 1$ and this together with the surjectivity of ${}^t\sigma$ contradicts the regularity of $t^2\beta$. The other steps in the regularity of our exact sequence are proved in a similar way, and are left to the reader.

We now forget about the duality hypothesis, and assume only that there is no torsion on C. We shall give arguments which could be applied (with suitable modifications) to arbitrary varieties. We need a series of lemmas. Throughout, we refer to our regular exact sequence, and assume that k is a field of definition for α and β. The hypothesis that C is without torsion will be used only in Lemma 5.

LEMMA 1. *Let W be a subvariety of C. Then the cycle $\beta^{-1}(W)$ is defined and contains one component with multiplicity* 1. *Its support consists of those points $b \in B$ such that $\beta(b) = W$.*

Proof: Let w be a generic point of C over k. Then by the definition of the factor group and F—VII$_6$ Th. 12 we see that

$$\Gamma_\beta \cdot (B \times w)$$

contains one component with multiplicity 1. We can make a translation on $B \times C$ which leaves Γ_β invariant, and moves $B \times w$ into $B \times w'$, with an arbitrary point w' of C. This shows that $\Gamma_\beta \cdot (B \times w')$ contains one component with multiplicity 1. We reduce our lemma to the case where W is a point by taking generic hyperplane sections. We know that C has a projective embedding. Let H be a generic hyperplane section of C. Then we can use associativity on the cycle

$$\Gamma_\beta \cdot (B \times W) \cdot (B \times H),$$

thus showing that this cycle is equal to $\Gamma_\beta \cdot (B \times (W \cdot H))$. Since $W \cdot H$ is a variety with multiplicity 1, we see by induction that $\beta^{-1}(W)$ has only one component with multiplicity 1. Our last assertion is obvious.

LEMMA 2. *Let X be a cycle on B and Y a cycle on C such that $X = \beta^{-1}(Y)$. Then Y is uniquely determined by this condition. In other words, if $\beta^{-1}(Y) = 0$, then $Y = 0$. If X is rational over a field K, then Y is rational over K. If X is a variety, so is Y, and it is equal to the set-theoretic image $\beta(X)$.*

Proof: Let Y_1, Y_2 be two subvarieties of C such that the algebraic sets $\beta^{-1}(Y_1)$ and $\beta^{-1}(Y_2)$ are equal. If we apply β set-theoretically, we obtain Y_1, Y_2 back. This shows that our cycle Y in the first assertion is uniquely determined. Thus β^{-1} gives an isomorphism of the group of cycles on C into that of B. If X is a variety, defined over K, then so is Y, namely if x is a generic point of X over k, then $\beta(x) = y$ is one for Y over K. If X is defined over an algebraic extension of K, and σ is an automorphism of the universal domain leaving K fixed, then $X^\sigma = \beta^{-1}(Y^\sigma)$. Moreover, the inseparability degree of Y over K is at most equal to that of X over K. Hence if X is rational over K, then the cycle Y of C such that $X = \beta^{-1}(Y)$ is also rational over K.

LEMMA 3. *Let X_1, X_2 be two subvarieties of B, of the same dimension, defined over K, and let v be a generic point of B over K. Then if the intersections $X_1 \cap A_v$ and $X_2 \cap A_v$ are not empty, and are equal, then $X_1 = X_2$. In addition, putting $X = X_1 = X_2$, every generic point of a component of $X \cap A_v$ over $K(v)$ is a generic point of X over K, and we have $\beta(X) = C$ (set-theoretically).*

Proof: From the theory of the factor group (Chapter I, § 1) the smallest field of definition of A_v containing K is equal to the field $K(w)$ with $w = \beta v$. If the intersection $X \cap A_v$ is not empty, its components are therefore algebraic over $K(w)$. Let x' be a generic point of one of the $K(w)$-components of this intersection, over $K(w)$. We have a specialization $v \to x'$ over $K(w)$, and *a fortiori* over K. Hence $w = \beta x'$. Let $\bar{x}$ be a generic point of X over K. By definition, we have a specialization $v \to \bar{x} \to x'$ over K. Put $\bar{w} = \beta\bar{x}$. Since $w = \beta x'$, it follows that $\bar{w}$ is a generic specialization of w over k. After a suitable isomorphism of $K(\bar{x})$ over K, we can therefore replace $\bar{x}$ by a generic point x of X over K, such that $w = \beta x$. We then see that x' is a specialization of x over $K(w)$, and hence that the locus of x over $K(w)$ is a $K(w)$-variety contained in $X \cap A_v$, and containing the $K(w)$-component of $X \cap A_v$ which is the locus of x' over $K(w)$. It follows that the specialization $x \to x'$ over $K(w)$ is actually an isomorphism, and we see that x' is a generic point of X over K. This proves the second assertion of our lemma. The first is an immediate consequence of it.

LEMMA 4. *Let X be a divisor on B, rational over K. Let v be a generic point of B over K. Then $X \cdot A_v = 0$ if and only if there exists a divisor Y on C such that $X = \beta^{-1}(Y)$.*

Proof: We may assume K algebraically closed. Suppose first that $X \cdot A_v = 0$. The components of X are of two different types: those whose intersection with A_v is empty, and those whose intersection with A_v is not. In the expression of X as a linear combination of varieties, let us separate these two types, and let us write $X = X' + X^*$ where X' contains only components of the first type and X^* components of the second type. The preceding lemma shows that we must have $X^* = 0$, i.e., the intersection of all components of X with A_v is empty. We may thus assume $X = X'$, and this brings us back to the case where X is a variety.

Consider the set-theoretic image $\beta(X)$ in C. It is a subvariety Y of C and we obviously have $\beta^{-1}(Y) \supset X$. The hypothesis $X \cap A_v$ empty shows that we cannot have $Y = C$. Hence $\beta^{-1}(Y)$ is not equal to B. Since X is a divisor on B, we must have $X = \beta^{-1}(Y)$.

Conversely, suppose $X = \beta^{-1}(Y)$ with some divisor Y on C.

Then $\beta(X) = Y$. The preceding lemma shows that $X \cap A_v$ is empty.

LEMMA 5. *Suppose that the torsion group on C is trivial. Let*

$$\beta^{-1} : N(C) \to N(B)$$

be the homomorphism induced on $D(C)/D_a(C)$ into $D(B)/D_a(B)$. Then β^{-1} is an injection.

Proof: By the complete reducibility theorem of Poincaré, there exists a homomorphism $\lambda : C \to B$ such that $\beta\lambda = n\delta_C$. We then have $(\beta\lambda)^{-1} = \lambda^{-1}\beta^{-1} = (n\delta)^{-1}$. But according to Proposition 2 of Chapter IV, § 1 we have $(n\delta)^{-1}(\eta) = n^2 \cdot \eta$ for $\eta \in N'(C)$. Hence the torsion group of C contains the kernel of β^{-1}. Since C is assumed to be without torsion, it follows that β^{-1} is injective.

Let us now return to the proof of Theorem 10. We consider first ${}^t\beta$. Let $Y \in D_a(C)$, and suppose $\beta^{-1}(Y) \sim 0$ on B. Let K be a field of rationality for Y containing k. There exists a function φ on B defined over K such that $\beta^{-1}(Y) = (\varphi)$. Let v be a generic point of B over K, and u a generic point of A over $K(v)$. If we make a translation on $B \times C$ by the point $(u, \beta u) = (u, 0)$ we see that the translate φ_u, i.e., the function such that $\varphi_u(v) = \varphi(v-u)$ has the same divisor as φ and hence that there exists a constant $c(u)$ in $K(u)$ such that $\varphi(v - u) = c(u)\varphi(v)$. The mapping $u \to c(u)$ is a generic homomorphism of A into the multiplicative group, which is therefore constant. This means that $c(u)$ is in K. Taking $u = 0$, we see that this can happen only if it is equal to 1. We see therefore that φ is invariant by generic translations of A. By the definition of the factor group, there exists a function ψ on C defined over K such that $\varphi(v) = \psi(\beta v)$, i.e., $\varphi = \psi \circ \beta$. According to Corollary 3 of Theorem 3 of the Appendix, § 1 we conclude that $\beta^{-1}[Y - (\psi)] = 0$. Lemma 2 gives $Y = (\psi)$. This proves that ${}^t\beta$ is injective.

Let us prove that ${}^t\beta$ is regular. Let X be a divisor on B, and in $D_a(B)$. Assume that $X \sim \beta^{-1}(Y)$ with $Y \in D_a(C)$. Let K be a field of rationality for X and v a generic point of B over K, and over a field of rationality for Y. We know from Proposition 4 of Chapter III, § 3 that $X_{-v} \sim X$. Since $\mathrm{Cl}(X)$ is in the image of

${}^t\beta$, its image under ${}^t\alpha$ must be 0. This means that $X_{-v} \cdot A$ is linearly equivalent to 0 on A. Translating by v, we see that $X \cdot A_v$ is linearly equivalent to 0 on A_v. Put $w = \beta v$. Since A_v is defined over $K(w)$, there exists a function θ on A_v, defined over $K(w)$, such that $(\theta) = X \cdot A_v$. There exists a function φ on B, defined over K, such that $\varphi(v) = \theta(v)$. In fact, θ is the function on A_v induced by φ. By Proposition 3 of Chapter III, § 1 we see that $[X - (\varphi)] \cdot A_v = 0$. According to Lemma 4, there exists a divisor Y_1 on C such that $X - (\varphi) = \beta^{-1}(Y_1)$ and Y_1 is rational over K, according to Lemma 2. Furthermore, $\beta^{-1}(Y) \sim \beta^{-1}(Y_1)$, and in particular, these two divisors are algebraically equivalent on B. Lemma 5 shows that Y_1 is algebraically equivalent to Y on C, and hence $Y_1 \in D_a(C)$. This proves that ${}^t\beta$ is regular because Y_1 is rational over K.

Finally, let us consider ${}^t\alpha$. Let $X \in D_a(B)$, and assume that $X \cdot A$ is defined and ~ 0 on A. As above, we get $X_{-v} \cdot A \sim 0$, and hence $X \cdot A_v \sim 0$ on A_v. The same argument that we already used shows that there exists a divisor $Y \in D_a(C)$ such that $X \sim \beta^{-1}(Y)$, and hence that the kernel of ${}^t\alpha$ is contained in the image of ${}^t\beta$. The converse was trivial, and this proves that the kernel of ${}^t\alpha$ is connected.

Actually, the duality hypothesis implies the lack of torsion. Indeed, using Corollary 2 of Theorem 4, § 3 we can assume that we have an exact sequence as in Theorem 10, where B has a trivial torsion group. Let Y be a divisor on C such that $nY \sim 0$. Then $\beta^{-1}(Y)$ is in $D_a(B)$. Furthermore, by Lemma 4, we know that

$$A_v \cdot \beta^{-1}(Y) = 0$$

for v generic. Making the translation $-(v, \beta v)$ on $B \times C$, we leave the graph of β invariant, and thus we get

$$A \cdot \beta^{-1}(Y_{\beta v}) = 0.$$

But $Y_{\beta v} \sim Y$ by Proposition 4 of Chapter III, § 3. Hence $\mathrm{Cl}[\beta^{-1}(Y)]$ is contained in the kernel of ${}^t\alpha$. Under the duality hypothesis, it is in the image of ${}^t\beta$, in other words, there exists a divisor $Y_1 \in D_a(C)$ such that $\beta^{-1}(Y_1) \sim \beta^{-1}(Y)$, and hence $\beta^{-1}(Y_1 - Y) \sim 0$. Furthermore, since ${}^t\beta$ is injective, it follows that

$pY_1 \sim 0$, and thus $p(Y_1-Y) \sim 0$. By Lemma 2 of Chapter VI, § 3 it follows that $Y_1-Y \sim 0$, and hence that Y is in $D_a(C)$. Summarizing:

THEOREM 10. *Under the duality hypothesis, the torsion group of an abelian variety is trivial.*

§ 5. Duality between image and trace

We can complete the theory of the K/k-image and trace by duality statements.

THEOREM 11. *Let K be a primary extension of k, and let A be an abelian variety defined over K. Assume the duality hypothesis. Then (B, τ) is a K/k-trace of A if and only if $(\hat{B}, {}^t\tau)$ is a K/k-image of $\hat{A}$.*

Proof: By symmetry we may consider the implication in one direction only, say (B, τ) is a K/k-trace of A. Given a homomorphism of $\hat{A}$ into an abelian variety C_1, we can use the duality hypothesis to write $C_1 = \hat{C}$ for some C, identifying $\hat{\hat{C}}$ with C under κ_C. Let $\gamma_1 : \hat{A} \to \hat{C}$ be a homomorphism, with γ_1 defined over K and $\hat{C}$ defined over k. Using Proposition 8 of Chapter V, § 2, and our hypothesis that κ_C and κ_A are birational, we see that there exists a homomorphism $\gamma : C \to A$ defined over K such that $\gamma_1 = {}^t\gamma$, namely

$$\gamma = \kappa_A^{-1}\, {}^t\gamma_1\, \kappa_C .$$

We can factor γ through (B, τ), and taking the transpose shows that we can factor ${}^t\gamma$ through $(\hat{B}, {}^t\tau)$. This proves what we wanted.

The proof of the regularity theorem for the K/k-trace was much easier than for the K/k-image. Hence if one has the duality hypothesis, one may omit completely the elaborate arguments of Theorem 3, and deduce this theorem from Theorem 9, by means of the following result.

PROPOSITION 8. *Let $k(u)$ be a finitely generated regular extension of k and A_u an abelian variety defined over $k(u)$. Let (B, τ_u) be its $k(u)/k$-trace. Let $u_1, \ldots, u_m$ be independent generic specializations of u over k, and let T_m be given by*

$$T_m(y) = (\tau_{u_1} y, \ldots, \tau_{u_m} y).$$

Then

$${}^tT_m : \hat{A}_{u_1} \times \ldots \times \hat{A}_{u_m} \to \hat{B}$$

is given by the formula

$${}^tT_m(x_1, \ldots, x_m) = \sum_{i=1}^{m} {}^t\tau_{u_i}(x_i).$$

Proof: We view τ_{u_i} as mapping B into the product, namely $\tau_{u_i}(y) = (0, \ldots, 0, \tau_{u_i} y, 0, \ldots, 0)$. Then $T_m = \sum \tau_{u_i}$ and taking the transpose using the linearity (Chapter V, § 1) we get what we want.

Of course, the proposition has an analogue in the opposite direction. We leave its formulation to the reader. When we need it or its dual, we shall refer merely to this proposition.

As a last application of the duality between image and trace, we give the dual of Theorem 4.

THEOREM 12. *Let the notations be as in Theorem* 4, § 2. *Then* $(\hat{A}, {}^ti_*)$ *is a* $k(u)/k$*-trace of* $\hat{B}_u$ *(under the assumption that abelian varieties have no torsion).*

Proof: By Theorem 4 and the regularity theorem, we can apply Proposition 8 and Theorem 9 to get our result. Of course, under the duality hypothesis, we can apply Theorem 11 directly.

[Added 1983 for the Springer edition: a proof of the duality hypothesis and of the fact that there is no torsion can be found in Mumford's book.]

§ 6. *Exact sequences of varieties*

We shall give only results concerning the Albanese variety. For more complete results, we refer the reader to Chow [15], who also gets more precise statements for the Picard variety.

Let $f : U \to V$ be a generically surjective rational map, defined over an algebraically closed field k. Let v be a generic point of V over k, and u a generic point of U such that $f(u) = v$. We shall assume that $k(u)$ is regular over $k(v)$, or in other words that f is regular. The locus of u over $k(v)$ is then a variety $W_v = W$ which is none other than $f^{-1}(v)$ in the sense of F—VII_6 Th. 12. We have the inclusion

$$i : W \to U$$

and fi is constant, because $fi(u) = v$ is constant relative to the field $k(v)$. Thus we have a sequence of rational maps

$$W \xrightarrow{i} U \xrightarrow{f} V$$

aud we shall say that this sequence is *generically exact.*

THEOREM 13. *Let*

$$W \xrightarrow{i} U \xrightarrow{f} V$$

be a generically exact sequence, relative to the generically surjective rational map $f : U \to V$. *Let* $f_* : A(U) \to A(V)$ *and* $i_* : A(W) \to A(U)$ *be the induced homomorphisms on the Albanese varieties. Let* $\lambda : A(W) \to A_0$ *be the canonical homomorphism of* $A(W)$ *onto its* $k(v)/k$*-image* A_0, *and let* $\alpha : A_0 \to A(U)$ *be the homomorphism which makes the following diagram commutative:*

$$\begin{array}{ccc} A(W) & & \\ \downarrow \lambda & \searrow^{i_*} & \\ A_0 & \xrightarrow{\alpha} & A(U) \end{array}$$

Then the sequence

$$0 \to A_0 \xrightarrow{\alpha} A(U) \xrightarrow{f_*} A(V) \to 0$$

is exact, up to isogenies.

Proof: Note first that

$$0 = (fi)_* = f_* i_*$$

because fi is constant. Hence the image of α is contained in the kernel of f_*.

In our statement of the theorem, we have used the Albanese variety $A(W)$ of W, rather than its $k(v)$-Albanese variety. We know that these differ by a purely inseparable isogeny. By making a purely inseparable extension of $k(v)$, we obtain a purely inseparable extension of $k(u)$ which changes our Albanese varieties only by purely inseparable isogenies. Since we have stated our theorem only up to isogenies, it follows that we may assume that the $k(v)$-Albanese variety of W_v is actually its Albanese variety.

To go further, we construct an approximate section of V in U, similar to that used in the proof of Poincaré's complete reducibility theorem. As a consequence, we shall obtain a map of W into its $k(v)$-Albanese variety equal to m times the canonical map.

Let P be a simple point of W, separable algebraic over $k(v)$ and such that f is defined at P. It suffices to take for P a sufficiently general specialization of u over $k(v)$. Let P_i $(i = 1, \ldots, d)$ be its conjugates over $k(v)$. We have a canonical admissible map

$$\varphi : W \times W \to A(W)$$

defined over $k(v)$. The cycle $\sum P_i$ is rational over $k(v)$, and hence over $k(u)$. The point $\sum \varphi(u, P_i)$ is rational over $k(u)$, and gives a rational map of W into $A(W)$ defined over $k(v)$. If we write φ over the algebraic closure of $k(v)$ as a difference of two maps of W into $A(W)$, then one sees immediately that our rational map is equal to $m \cdot \varphi_W$ where $\varphi_W : W \to A(W)$ is a canonical map of W into $A(W)$. Note that φ_W is not necessarily defined over $k(v)$, but that $m \cdot \varphi_W$ is defined over $k(v)$. Hence the induced homomorphism

$$m \cdot \varphi_{W*} : A(W) \to A(W)$$

is equal to $m \cdot \delta_{A(W)}$.

We can define a rational map $g : U \to A_0$ by the formula $g(u) = \lambda m \cdot \varphi_W(u)$, and the following diagram is commutative:

$$\begin{array}{ccc} W & \xrightarrow{\ i\ } & U \\ {\scriptstyle m \cdot \varphi_W}\downarrow & & \downarrow{\scriptstyle g} \\ A(W) & \xrightarrow[\ \lambda\ ]{} & A_0 \end{array}$$

that is to say, $gi = m \cdot \lambda\varphi_W$. Denoting by a $*$ the homomorphisms induced on the Albanese varieties, we find

$$g_* i_* = m \cdot \lambda\delta_{A(W)}.$$

I contend that $g_* \alpha = m \cdot \delta_{A_0}$. Indeed, we have

$$g_* \alpha\lambda = g_* i_* = m \cdot \lambda\delta_{A(W)} = m \cdot \delta_{A_0}\lambda.$$

Since λ is surjective, this implies that $g_* \alpha = m \cdot \delta_{A_0}$, and shows that α is an isogeny.

We are now going to show that the product mapping

$$(g_*, f_*) : A \to A_0 \times A(V)$$

is an isogeny.

In the first place, we can define a rational map

$$h : V \to A(U)$$

as follows: Let $\varphi_U : U \to A(U)$ be a canonical map of U into its Albanese variety, defined over k. The point $\sum \varphi_U(P_i)$ is rational over $k(v)$, and we get h by the formula

$$h(v) = \sum \varphi_U(P_i).$$

Let $h_* : A(V) \to A(U)$ be the induced homomorphism.

Observe that $f_* i_* = 0$ gives $f_* \alpha = 0$. I contend that

$$f_* h_* = m \cdot \delta_{A(V)}.$$

Indeed, we have

$$f_* h(v) = \sum f_* \varphi_U(P_i) = \sum \varphi_V f(P_i) = m \cdot \varphi_V(v) + c$$

where c is a constant, this being due to the fact that $f(P_i) = v$ and also to the commutative diagram (up to an additive constant):

$$\begin{array}{ccc} U & \xrightarrow{\varphi_U} & A(U) \\ {\scriptstyle f}\downarrow & & \downarrow{\scriptstyle f_*} \\ V & \xrightarrow[\varphi_V]{} & A(V) \end{array}$$

If we now take the induced homomorphisms on the Albanese varieties, we get

$$f_* h_* = m \cdot \varphi_{V*} = m \cdot \delta_{A(V)}.$$

We have thus obtained homomorphisms which can be described briefly by the following diagram:

$$A_0 \underset{g_*}{\overset{\alpha}{\rightleftarrows}} A(U) \underset{h_*}{\overset{f_*}{\rightleftarrows}} A(V)$$

and we know that $g_* \alpha = m \cdot \delta_{A_0}$ and $f_* h_* = m \cdot \delta_{A(V)}$. In addition we have seen that $f_* \alpha = 0$.

From all this, we deduce that the intersection $\alpha A_0 \cap h_* A(V)$ is finite. Hence the connected component of the kernel of the restriction of f_* to $\alpha A_0 + h_* A(V)$ can only be αA_0, and the connected component of the kernel of (g_*, h_*) must be contained in αA_0. Since $g_* \alpha = m \cdot \delta_{A_0}$, it must be equal to 0. This proves that the product mapping

$$(g_*, f_*) : A(U) \to A_0 \times A(V)$$

is surjective.

To conclude the proof, there remains but to show that $\dim A \leqq \dim A_0 + \dim A(V)$. For this, we note that the commutative diagram (up to a constant)

$$\begin{array}{ccc} W & \xrightarrow{i} & U \\ {\scriptstyle m\cdot\varphi_W}\downarrow & & \downarrow{\scriptstyle m\cdot\varphi_U} \\ A(W) & \xrightarrow[i_*]{} & A(U) \end{array}$$

gives us

$$m \cdot \varphi_U(u) = \alpha\lambda m \cdot \varphi_W(u) + c.$$

Since $m \cdot \varphi_W$ is defined over $k(v)$, and $m \cdot \varphi_U$ over k, it follows that the constant c is rational over $k(v)$. This shows that the map ψ such that

$$\psi(u) = m \cdot \varphi_U(u) - \alpha g(u),$$

can be factorized through V, i.e., that we can write

$$\psi(u) = \beta\varphi_V f(u).$$

If we take the induced homomorphism, we find

$$m \cdot \delta_{A(U)} = \alpha g_* + \beta f_*.$$

This proves the desired inequality, and concludes the proof.

REMARK. In certain cases, it is convenient to have a criterion for $A(V)$ to be trivial. The variety $W = W_v$ can be viewed as a generic variety in an algebraic system, which is called a *fiber system* of U, with parameter variety V. A base point of this system is a point in all the members of the system. Such a point has the property that

$$u \to P$$

is a specialization over $k(v)$, and that P is rational over k. If P is simple on U, then one sees immediately that $A(V) = 0$ because the rational map of U into $A(V)$ induced by that of V cannot be defined at P, and we use Theorem 2 of Chapter II, § 1.

Historical note:

This entire chapter is due to Chow [14], [15]. The dual of Lefschetz's theorem concerning the Picard variety of a generic hyperplane section had been considered for a non-singular variety by Matsusaka [58], but one should note that in this article, an analysis of the proofs shows that in fact the isomorphism obtained may be purely inseparable.

As mentioned before, a propos of the field of definition of the Albanese variety, we have used here for the analogous problem of the K/k-image the criteria of Chow in the form given to them by Weil [92].

Finally (added in proof) we mention that the duality hypothesis has recently been proved by Cartier, thus validating the results in this chapter which have been proved under its assumption. In particular, we have formally $\hat{\hat{A}} = A$, and more precisely, if $(\hat{A}, D)$ is a Picard variety of A, then $(A, {}^tD)$ is a Picard variety of $\hat{A}$. A sketch of the proof can be found in the Bourbaki Seminar, 1957–58.

APPENDIX

Composition of Correspondences

We have collected here a number of auxiliary results which have been used throughout the book. The proofs are based exclusively on *Foundations.*

§ 1. *Inverse images*

In the first place, we have the following theorem, which is simply a restatement of F—VII_6 Th. 16 under less restrictive hypotheses.

THEOREM 1. *Let* U^n *be a variety,* V *a complete variety,* X^r *a cycle on* $U \times V$, *with* $r \leqq n$, *and* Y^s *a cycle on* U. *Assume that*

(a) $X \cdot (Y \times V)$ *is defined on* $U \times V$,

(b) *there is no subvariety* Z *singular on* $U \times V$, *contained in* supp $(X) \cap$ supp $(Y \times V)$ *whose projection on* U *is of dimension* $\geqq r + s - n$ *and simple on* U.

Then $(\mathrm{pr}_U X) \cdot Y$ *is defined on* U, *and we have*

$$\mathrm{pr}_U[X \cdot (Y \times V)] = (\mathrm{pr}_U X) \cdot Y.$$

Proof: It is the same as in *Foundations.*

We shall now consider the composition of correspondences on products of varieties. Let U, V, W be three varieties, X a cycle on $U \times V$, and Y a cycle on $V \times W$. If the intersection

$$(X \times W) \cdot (U \times Y)$$

is defined on $U \times V \times W$, we shall say that the cycle

$$\mathrm{pr}_{13}[(X \times W) \cdot (U \times Y)]$$

is the *composition*, or the cycle *composed of* X and Y, and we shall denote it by $Y \circ X$. We shall say that $Y \circ X$ is *defined* if the above intersection is defined.

A subvariety X of a product $U \times V$ will be called *degenerate* on U (resp. on V) if its projection on U (resp. on V) is not equal to U (resp. to V). The projection here is of course to be understood in the usual geometric sense: it is the closure under the Zariski topology of the set-theoretic projection. It is denoted by $\mathrm{proj}_U(X)$. We shall say that X is *non-degenerate* on U if its projection is not degenerate, i.e., if its projection on the first factor is U. If X is a cycle on $U \times V$, we say that X is non-degenerate on U if all its components are non-degenerate. We say that X is *degenerate* if it is degenerate on both U and V.

Let X be a divisor on $U \times V$. Suppose that X is degenerate on U, and suppose that X is a variety. Let $X' = \mathrm{proj}_U(X)$. Then $X \subset X' \times V$. Since these two subvarieties of $U \times V$ have the same dimension, it follows that $X = X' \times V$. Thus we see that a degenerate divisor on $U \times V$ is of type $X_1 \times V + U \times X_2$ where X_1 and X_2 are divisors on U and V, respectively.

Let Z be a cycle on $U \times V$, and X a cycle on U. We write

$$Z(X) = \mathrm{pr}_2[Z \cdot (X \times V)]$$

and, if Y is a cycle on V, we write

$${}^tZ(Y) = \mathrm{pr}_1[Z \cdot (U \times Y)]$$

whenever these intersections are defined. If Z is the graph of a rational map $f : U \to V$, then we also write

$${}^tZ(Y) = f^{-1}(Y).$$

Usually, we denote the graph of f by Γ_f.

Let f again be a rational map, and let U' be a subvariety of U along which f is defined. Let k be a field of definition for f, U, V, and U', and let u' be a generic point of U' over k. Then $f(u')$ is the generic point of a subvariety of V. We shall denote it by $f(U')$, and call in the *set-theoretic image* of U'. It is in general distinct from the cycle $\Gamma_f(U') = \mathrm{pr}_2[\Gamma_f \cdot (U' \times V)]$, even if $U' = U$, because the dimension of $f(U)$ may be lower than that of U. The cycle $\Gamma_f(U')$ will sometimes be denoted also by $f(U')$, and we shall always specify in this case that we take the image

$f(U')$ *in the sense of intersection theory*. The context will usually make our meaning clear.

Except in Theorem 5 below, we shall deal only with the set-theoretic image for this entire section.

We shall study the composition of rational maps, and shall work with the following hypotheses.

Basic hypotheses: Let U^n, V, W be three varieties, and assume that V and W are complete. Let

$$f : U \to V, \quad \text{and} \quad g : V \to W$$

be two rational maps such that $f(U)$ is simple on V, g is defined at $f(U)$, and $g(f(U))$ is simple on W. Let Γ_f and Γ_g be the graphs of f and g respectively. We then have a rational map $h : U \to W$ composed of f and g, and noted by $g \circ f$. If u is a generic point of U over a field of definition k for f and g, we denote by Γ the locus of $(u, f(u), g(f(u)))$. We have $\mathrm{pr}_{13}\, \Gamma = \Gamma_h$. Finally, we let Z be a cycle on W, and we assume that $g^{-1}(Z)$ is defined.

PROPOSITION 1. *Assume the basic hypotheses. Then Γ is a component of $(\Gamma_f \times W) \cap (U \times \Gamma_g)$ and is the only component whose projection on U is U. If furthermore $(\Gamma_f \times W) \cdot (U \times \Gamma_g)$ is defined, then this cycle is equal to $\Gamma + X$, where X is a positive cycle such that* $\mathrm{proj}_U X \neq U$.

Proof: If (u, v, w) is a point of the intersection such that u is generic on U, then we have necessarily $v = f(u)$ and $w = h(u)$. Hence Γ is a component of the intersection, and is the only component whose projection on U is U. Assume now that the intersection product is defined. We must show that Γ appears with multiplicity 1. Let u be again a generic point of U, and consider the intersection

$$(u \times V \times W) \cdot (\Gamma_f \times W) \cdot (U \times \Gamma_g).$$

One sees immediately that we can apply the associativity theorem. On one hand this cycle is equal to

$$(u \times f(u) \times W) \cdot (U \times \Gamma_g) = (u \times f(u) \times h(u)),$$

and on the other hand, we obtain

$$(u \times V \times W) \cdot (\Gamma + X) = (u \times V \times W) \cdot \Gamma.$$

We see therefore, that the multiplicity of Γ must be equal to 1, thus proving the proposition.

THEOREM 2. *Assume the basic hypotheses, and in addition that V, W are non-singular, that f, g are everywhere defined, and that $h^{-1}(Z)$ is defined. Then $f^{-1}(g^{-1}(Z))$ is also defined, and*

$$f^{-1}(g^{-1}(Z)) = h^{-1}(Z).$$

Proof: In Proposition 1, Γ is the only component of intersection because f and g are everywhere defined. This hypothesis also shows that the composed cycle of Γ_f and Γ_g is defined. We shall now make a list of the formal steps of the proof, and we shall justify them afterwards.

$$\begin{aligned}
h^{-1}(Z) &= \mathrm{pr}_1\,[\Gamma_h \cdot (U \times Z)] && (1)\\
&= \mathrm{pr}_1\,\{\mathrm{pr}_{13}\,[(\Gamma_f \times W) \cdot (U \times \Gamma_g)] \cdot (U \times Z)\} && (2)\\
&= \mathrm{pr}_1\mathrm{pr}_{13}\,\{[(\Gamma_f \times W) \cdot (U \times \Gamma_g)] \cdot (U \times V \times Z)\} && (3)\\
&= \mathrm{pr}_1\,\{(\Gamma_f \times W) \cdot [U \times (\Gamma_g \cdot (V \times Z))]\} && (4)\\
&= \mathrm{pr}_1\mathrm{pr}_{12}\,\{[(\Gamma_f \times W) \cdot [U \times (\Gamma_g \cdot (V \times Z))]\} && (5)\\
&= \mathrm{pr}_1\{\Gamma_f \cdot [U \times g^{-1}(Z)]\} && (6)\\
&= f^{-1}(g^{-1}(Z)). && (7)
\end{aligned}$$

Step (1) comes from the definitions and assumptions.

Step (2) consists in substituting $\mathrm{pr}_{13}\,\Gamma$ for Γ_h.

Step (3) consists in applying the projection theorem with respect to V. It is valid if the intersection in $\{\ \}$ is defined. This comes from the assumptions that f, g are everywhere defined and that $h^{-1}(Z)$ is defined.

Step (4) accomplishes two things. First we take off pr_{13} by F—VII$_6$ Th. 14 (ii). Next, the hypothesis that $g^{-1}(Z)$ is defined allows us to use associativity, which gives us the expression in (4).

Step (5) adds pr_{12}.

Step (6) uses the projection theorem with respect to W, and by F—VII$_6$ Th. 14 (iii) we can separate pr_{12} on U and on $\Gamma_g \cdot (V \times Z)$ to obtain what we want.

Step (7) consists in applying the definition of f^{-1}.

This concludes the proof.

We remark particularly that the projection theorem was used crucially with respect to V and W, and that we needed the fact that they are non-singular. We shall see later other cases where we can do away with this assumption.

PROPOSITION 2. *In addition to the basic hypotheses, assume that W is a complete non-singular curve. Let Z be a point on W, and assume that there is no subvariety U'^{n-1} of U, simple on U, such that $f(U')$ is singular on V, h is constant on U', and $h(U') = Z$. Then*

$$(\Gamma_f \times W) \cdot (U \times \Gamma_g)$$

is defined, and is of type $\Gamma + T \times W$ where T is a cycle of dimension $n - 1$ on $U \times V$, whose support is contained in Γ_f. The expressions $h^{-1}(Z)$ and $f^{-1}(g^{-1}(Z))$ are defined, and we have

$$f^{-1}(g^{-1}(Z)) = h^{-1}(Z) + \mathrm{pr}_U T.$$

Proof: Note first that the statement of our proposition depends on the fact that V and W are complete, and f, h are therefore defined on every subvariety of U of codimension 1.

We observe that $(U \times \Gamma_g)$ is a divisor on $U \times V \times W$ and that our hypotheses imply that $(\Gamma_f \times W)$ is not contained in this divisor. Hence the intersection product of these cycles is defined by F—VII_6 Prop. 16.

According to Proposition 1, we know that all the components of the cycle X in this proposition have a projection on U of dimension lower than that of U, that is to say, X is degenerate on U. We have already noted that we can then write $X = T \times W$ with a suitable divisor T on $U \times V$.

Consider the intersection

$$(\Gamma_f \times W) \cdot (U \times \Gamma_g) \cdot (U \times V \times Z).$$

Since $U \times V \times Z$ is a divisor on $U \times V \times W$, our hypotheses imply that this intersection is defined and that we can apply the associativity theorem.

We now begin with step (3) in the proof of Theorem 2. We

note that steps (4), (5), (6), and (7) are valid in the present situation because we have used the projection theorem in these steps only with respect to W which is still complete and non-singular in the present case.

We must therefore verify that we can retrace our steps back from step (3) to step (1), and for this we must project with respect to V. We must thus verify that the hypotheses of Theorem 1 are satisfied. This is trivial, and when we take

$$\mathrm{pr}_{13}\,[(\Gamma_f \times W)\cdot(U \times \Gamma_g)]$$

we obtain $\Gamma_h + (\mathrm{pr}_1\, T \times W)$. We can now go trivially from step (2) to step (1), and our proposition is proved.

THEOREM 3. *In addition to the basic hypotheses, assume that W is a complete non-singular curve, and that Z is a cycle of dimension 0 and degree 0 on W. Assume also that there is no simple subvariety U'^{n-1} of U such that $f(U')$ is singular on V, h is constant on U', and $h(U')$ is a component of Z. Then $h^{-1}(Z)$ and $f^{-1}(g^{-1}(Z))$ are defined, and they are equal.*

Proof: This is an immediate consequence of Proposition 2, in view of the fact that the terms containing $\mathrm{pr}_U T$ will cancel because of our assumption that Z is of degree 0.

COROLLARY 1. *In addition to the basic hypotheses, assume that W is a curve, that V, W are non-singular, and that Z is a cycle of dimension 0 and degree 0 on W. Then $h^{-1}(Z)$ and $f^{-1}(g^{-1}(Z))$ are defined, and they are equal.*

COROLLARY 2. *Let $f: U \to V$ be a rational map which is generically surjective. Assume that V is complete, that U, V have the same dimension, are non-singular in codimension 1, and that neither f nor f^{-1} have fundamental points. Let Y be a divisor on V such that $Y \sim 0$ on V. Then $f^{-1}(Y)$ is defined and is ~ 0 on U.*

Proof: Take for W the projective straight line, and $Z = (0) - (\infty)$.

COROLLARY 3. *Let $f: U \to V$ be a rational map, and assume that V is complete, non-singular. Let Y be a divisor on V, $Y \sim 0$ on V. If $f^{-1}(Y)$ is defined, then $f^{-1}(Y) \sim 0$ on U.*

Proof: If Y is the divisor of the function $g : V \to W$ with W being the projective line, then our hypothesis that $f^{-1}(Y)$ is defined implies that $f(U)$ is not contained in the support of Y. Since V is non-singular, it is normal. Since g cannot have zeros or poles through $f(U)$, it follows by IAG—VI, Prop. 3 that g is defined at $f(U)$, and that its value at $f(U)$ cannot be either 0 or ∞. If we take $Z = (0) - (\infty)$, we see that we can apply Theorem 3.

The above corollary ends our study of the formalism concerning $f^{-1}(g^{-1}(Z))$, and we shall not make use any more of the basic hypotheses.

THEOREM 4. *Let $f : U \to V$ be an everywhere defined rational map, and assume that U, V are non-singular. Let Y, Z be two cycles on V. Assume that $Y \cdot Z$ is defined on V, and also that $f^{-1}(Y \cdot Z)$, $f^{-1}(Y)$, $f^{-1}(Z)$ are defined. Then $f^{-1}(Y) \cdot f^{-1}(Z)$ is defined on U, and we have*

$$f^{-1}(Y) \cdot f^{-1}(Z) = f^{-1}(Y \cdot Z).$$

Proof: By definition, we have

$$\begin{aligned} f^{-1}(Y \cdot Z) &= \mathrm{pr}_U\{\Gamma_f \cdot (U \times (Y \cdot Z))\} \\ &= \mathrm{pr}_U\{\Gamma_f \cdot [(U \times Y) \cdot (U \times Z)]\}. \end{aligned}$$

Using the associativity theorem, which is clearly applicable, we find that this expression is

$$= \mathrm{pr}_U\{[\Gamma_f \cdot (U \times Y)] \cdot (U \times Z)\}.$$

We shall now use F—VII_6 Th. 18, corollary. On the ambient variety $U \times V$, we consider the graph Γ_f as a subvariety on which we induce our intersections. According to the above reference, the last expression which we have obtained is equal to the following intersection, taken on Γ_f:

$$\mathrm{pr}_U\{[\Gamma_f \cdot (U \times Y)] \cdot [\Gamma_f \cdot (U \times Z)]\}$$

and in view of the fact that f is everywhere defined, and hence that pr_U gives a biholomorphic map between U and the graph of f, this is equal to the intersection

$$\mathrm{pr}_U[\Gamma_f \cdot (U \times Y)] \cdot \mathrm{pr}_U[\Gamma_f \cdot (U \times Z)]$$

taken on U. We have thus proved our theorem.

In the next theorem, the hypotheses essentially mean that U is a covering of V locally at Y.

THEOREM 5. *Let $f: U \to V$ be a generically surjective rational map, and assume that U, V have the same dimension. Let d be the degree of f. Let Y be a cycle on V and assume:*

(a) *that the graph Γ_f of f lies properly above each component of Y;*

(b) *that all the subvarieties of Γ_f above components of Y are simple on U;*

(c) *that f is defined at every component of $f^{-1}(Y)$. Then we have*

$$f(f^{-1}(Y)) = d \cdot Y,$$

taking f in the sense of intersection theory.

Proof: By linearity, we may assume that Y is a subvariety of V. Let Y_i^* be the subvarieties of Γ_f whose projection on the second factor is Y. Then

$$\Gamma_f \cdot (U \times Y) = \sum e_i Y_i^*$$

with suitable multiplicities e_i. If X_i is the projection of Y_i^* on U, then using our assumption that f is defined along the X_i, we find

$$\Gamma_f \cdot (X_i \times V) = Y_i^*.$$

We can now apply F—VII$_5$ Th. 8, observing that the projection of Γ_f on V is equal to V by assumption, and therefore that

$$[\Gamma : \Gamma'] = d.$$

§ 2. Divisorial correspondences

We are going to be particularly interested in taking inverse images of divisors. Since we shall work modulo linear equivalence, we shall now discuss the possibility of defining such inverse images for linear equivalence classes.

According to Zariski ("The concept of a simple point on an

algebraic variety," Trans. Amer. Math. Soc., July 1947, Vol. 62, No. 1, pp. 1—52), one knows that the local ring of a simple point on a variety is a unique factorization domain. The irreducible elements of this ring are in one-one correspondence with the divisors passing through this point. More precisely, suppose that V is defined over k, and let P be a simple point of V, algebraic over k. Let $\mathfrak{p}$ be the prime rational cycle determined by P over k. Let $\mathfrak{o}$ be the local ring of P in $k(V)$. Let φ be a function in $\mathfrak{o}$, which is a prime element of $\mathfrak{o}$. Then $(\varphi)_0$ can be written

$$(\varphi)_0 = X + Y$$

where X is a prime rational divisor over k, containing $\mathfrak{p}$, and Y is a divisor such that $P \notin \operatorname{supp}(Y)$.

Consequently, if X is an abitrary divisor on V, rational over k, there exists a function on V, defined over k, such that $X = (\varphi) + Y$, with a divisor Y such that $P \notin \operatorname{supp}(Y)$. We say that X is *locally linearly equivalent to* 0 at the simple point P of V.

Let W be a subvariety of V, defined over $\bar{k}$. Then W has a point P algebraic over k. Let X be a divisor on V, rational over k. From what we have seen above, there exists a function φ defined over k such that if we put $X_1 = X + (\varphi)$, then P, and hence W, are not contained in the support of X_1. Using F—VII$_6$ Prop. 16 we get:

PROPOSITION 3. *Let V be a variety, W a subvariety both defined over k. Let X be a divisor on V, rational over k. Then there exists a function φ on V, defined over k, such that if we put $X_1 = X + (\varphi)$, then $X_1 \cdot W$ is defined.*

COROLLARY. *Let X be a divisor on a product $U \times V$. Let k be a field of definition for U and V over which X is rational. Let P be a simple point of U. Then there exists a function φ on $U \times V$ defined over k, such that if we put $X_1 = X + (\varphi)$, then $X_1(P)$ is defined.*

Proof: Let U' be the locus of P over k. It is a prime rational cycle, and all the components of $U' \times V$ are simple on $U \times V$

and are defined over $\bar{k}$. The existence of the function φ then follows as above.

More generally, we have an analogous result for inverse images.

PROPOSITION 4. *Let $f : U \to V$ be a rational map defined over k, and such that $f(U)$ is simple on V. Let Y be a divisor on V, rational over $\bar{k}$. Then there exists a function φ on V, defined over k, such that if we put $Y_1 = Y + (\varphi)$, then $f^{-1}(Y_1)$ is defined.*

Proof: According to F—VII_6 Prop. 16, in order that $f^{-1}(Y_1)$ be defined, it suffices that Γ_f should not be contained in $U \times Y$, that is to say that $f(U)$ should not be contained in supp (Y). Since we have assumed $f(U)$ simple on V, there exists a simple point P of V contained in $f(U)$ and algebraic over k. It suffices to apply the preceding remarks to conclude the proof of our proposition.

Taking into account Proposition 4 and Corollary 3 of Theorem 3, § 1, we see that we may define the inverse image of a linear equivalence class under the hypothesis of that corollary.

A divisor on a product $U \times V$ will be called a *divisorial correspondence.* Divisors of type

$$X \times V + U \times Y + (\varphi)$$

where X is a divisor on U, Y a divisor on V, and φ a function on the product, form a subgroup of the group of divisors $D(U \times V)$, which is called the group of *trivial correspondences.* The factor group will be called the group of *correspondence classes* on $U \times V$.

PROPOSITION 5. *Let $U \times V$ be a product of two varieties, and D a divisor on $U \times V$ which we may write in two ways as a trivial correspondence:*

$$D = X \times V + U \times Y + (\varphi) = X_1 \times V + U \times Y_1 + (\varphi_1).$$

Then $X \sim X_1$ and $Y \sim Y_1$, in other words the degenerate components of a trivial correspondence are uniquely determined up to a linear equivalence. If in addition v' is a simple point of V such that ${}^tD(v')$ is defined, then we have $X \sim {}^tD(v')$.

Proof: Let u be generic on U over a sufficiently large field k, and take $D(u)$. According to F—VIII_2 Th. 4, Cor. 1, we obtain

Y and Y_1 up to linear equivalence. Similarly, taking ${}^tD(v)$ with v generic on V, we obtain X and X_1. Finally, suppose ${}^tD(v')$ is defined. Let θ be a function on V such that v' is not contained in the support of $Y + (\theta)$, and let ω be a function on $U \times V$ such that $U \times v'$ is not contained in the support of $(\omega\varphi)$. Put $Y_2 = Y + (\theta)$, and $D_2 = X \times V + U \times Y_2 + (\omega\varphi)$. We have $(U \times Y_2) \cdot (U \times v') = 0$ and $\mathrm{pr}_1\,[(\omega\varphi) \cdot (U \times v')] \sim 0$. This gives us $X \sim {}^tD(v')$.

COROLLARY. *Let D_1, D_2 be two divisors on a product $U \times V$, and assume that D_1, D_2 are in the same correspondence class. Let P, Q be two simple points of U such that $D_i(P)$ and $D_i(Q)$ are defined for $i = 1, 2$. Then we have*

$$D_1(P) - D_1(Q) \sim D_2(P) - D_2(Q).$$

Proof: Put $D = D_1 - D_2$. Then $D(P)$ and $D(Q)$ are defined and by hypothesis, D is linearly equivalent to a degenerate divisor. We can apply the proposition to D, and our assertion is then a reformulation of the fact that $D(P) \sim D(Q)$.

PROPOSITION 6. *Let D be a divisor on $U \times V$, and let k be a field of definition for U, V over which D is rational. Let u be a generic point of U over k. If $D \cdot (u \times V) = 0$, then D is degenerate on U.*

Proof: F—VII_6 Th. 12.

The following theorem is known as the *seesaw principle.*

THEOREM 6. *Let U, V be two varieties, and assume that V is complete and without singularities of codimension 1. Let D be a divisor on $U \times V$, and k a field of definition for U, V over which D is rational. Let u be a generic point of U over k, and assume that there exists a divisor Y on V, rational over k, such that $D(u) \sim Y$. Then there exists a function φ on $U \times V$, defined over k, and a divisor X of U, rational over k, such that*

$$D = X \times V + U \times Y + (\varphi).$$

Proof: Put $D_1 = D - U \times Y$. Then $D_1(u)$ is linearly equivalent to 0 on V, and rational over $k(u)$. According to Cor. 1 of F—VIII_3 Th. 10, there exists a function φ_u on V, defined over $k(u)$, such that $D_1(u) = (\varphi_u)$. If (u, v) is a generic point of $U \times V$ over k, we can find a function φ on $U \times V$ such that $\varphi(u, v) = \varphi_u(v)$. We then have $[D_1 - (\varphi)] \cdot (u \times V) = 0$ by F—VIII$_2$ Th. 4, Cor. 1, and consequently there exists a divisor X on U such that $D_1 = (\varphi) + X \times V$. Since D_1 is rational over k, and φ is defined over k, it follows that X is rational over k, thus proving our theorem.

COROLLARY. *Let D be a trivial correspondence on a product $U \times V$, and assume that U is complete and non-singular in codimension* 1. *Let k be a field of definition for U, V over which D is rational. Assume that V has a simple point Q rational over k. Then we can write*

$$D = X \times V + U \times Y + (\varphi)$$

with divisors X, Y rational over k, and a function φ defined over k.

Proof: There exists a function φ on $U \times V$ defined over k, such that if we put $D_1 = D - (\varphi)$, the intersection $D_1 \cdot (U \times Q)$ is defined (corollary of Proposition 3). According to Proposition 5, the linear equivalence class ${}^tD_1(Q)$ is constant on U, and is the same as that of ${}^tD(v)$, with v generic on V. We can apply Theorem 6 in the opposite direction, because ${}^tD_1(Q)$ is rational over k, and we have assumed U complete and non-singular in codimension 1. This gives us the desired expression for D_1, and hence for D.

THEOREM 7. *Let U^n, V, W be three varieties and assume that V is complete, non-singular. Let Z^n be a cycle on $U \times V$, and Y a divisor on $V \times W$. Assume that*

$$(Z \times W) \cdot (U \times Y)$$

is defined, and put $D = \mathrm{pr}_{13}\,[(Z \times W) \cdot (U \times Y)]$. *Let k be a field of definition for U, V, W over which Z and Y are rational. If u is a generic point of U over k, then we have*

$$Y(Z(u)) = D(u) \tag{1}$$

and if w is a generic point of W over k, then we have

$$ {}^tZ({}^tY(w)) = {}^tD(u) \qquad ({}^t1)$$

in the sense that the expressions which occur in these formulas are defined, and the terms on the left and the right of (1) *and* (^{t}1) *are equal.*

Proof: Consider the intersection

$$(u \times V \times W) \cdot (Z \times W) \cdot (U \times Y).$$

The two intersection products taken separately are defined. If (u, v, w) is a point in some component of $(u \times V \times W) \cap (Z \times W)$, then (u, v) is a generic point of a component of Z and w is a generic point of W independent of (u, v). By assumption, (u, v, w) is not contained in $U \times Y$. By F—VII$_6$ Props. 16 and 17 it follows that the three cycles intersect properly, and that we have associativity. If we take the projection with respect to V, we obtain

$$\mathrm{pr}_{13}\{(u \times V \times W) \cdot [(Z \times W) \cdot (U \times Y)]\} = (u \times W) \cdot D = u \times D(u).$$

On the other hand, by associativity, we also have

$$\begin{aligned} &\mathrm{pr}_{13}\{[(u \times V \times W) \cdot (Z \times W)] \cdot (U \times Y)\} \\ &\quad = \mathrm{pr}_{13}\{[u \times Z(u) \times W] \cdot (U \times Y)\} \\ &\quad = \mathrm{pr}_{13}\{u \times [(Z(u) \times W) \cdot Y]\} \\ &\quad = u \times \mathrm{pr}_W [(Z(u) \times W) \cdot Y] \\ &\quad = u \times Y(Z(u)). \end{aligned}$$

This proves formula (1), and its transpose is proved in the same manner, by considering the intersection

$$(Z \times W) \cdot (U \times Y) \cdot (U \times V \times w)$$

with w generic on W. We leave this to the reader.

Under the hypotheses of Theorem 7, we can then form the composed divisor $Y \circ Z$ on $U \times W$. If Z is the graph Γ_f of a rational map f, then we shall also write $Y \circ f$ for $Y \circ \Gamma_f$.

If at the beginning of the statement of Theorem 7 we did not assume that $Y \circ Z$ is defined, then the remarks show that we can find a divisor Y_1 on $V \times W$, also rational over k, and linearly equivalent to Y over k such that $Y_1 \circ Z$ is defined. In all our applications, we are interested in divisors only up to linear equivalence, and thus we shall always be able to assume that our composed divisor is defined.

BIBLIOGRAPHY

Before proceeding with our list of references, it is appropriate to make some further historical remarks, and to comment on our choice of articles.

On the whole, the bibliography contains only articles written since *Foundations*. The exceptions bear on certain special subjects which are not yet completely absorbed by the general theory of abelian varieties, for instance Deuring's results on the ring of endomorphisms of an elliptic curve.

We omit all references to topological or transcendental papers, partly for lack of competence, partly because such an enterprise would take us far away from the methods which we have used in this book, and partly because the reader will find excellent accounts of the topological and transcendental results in F. Conforto, *Abelsche Funktionen*, Springer, Berlin, **1956** and in M. Baldassari, Algebraic Varieties, *Ergebnisse der Mathematik*, Springer, Berlin, **1956**.

This latter volume also contains an excellent bibliography, including the superb work of the Italian school, and especially that of Castelnuovo and Severi for the topics which have interested us in this book. In particular, the reader may consult F. Severi, *Memorie Scelte*, Zuffini, Bologna, **1950**, and G. Castelnuovo, *Memorie Scelte*, Zanichelli, Bologna, **1937**, this latter collection including for instance "Sulle funzione abeliane," pp. **529—549** where the reader will find our divisor Θ and an analysis of its numerical properties (from which our notation has been taken).

Of course, the umbilical cord which tied algebraic geometers to the classical schools (mainly the Italian school) for so long, has now been cut, for better or for worse, by men like Weil and Zariski, in a large number of subjects. Not all, however, and the reader will find, for example, at the end of F. Severi, *Vorlesungen über Algebraische Geometrie*, Teubner, Leipzig, **1921** a thorough discussion of the problem of moduli, and the classification of algebraic systems of curves which is still far from having been absorbed by abstract algebraic geometry. To pick another example, as recently as three years ago, Weil could profitably extract an idea from S. Lefschetz, "On certain numerical invariants of algebraic varieties," Trans. Amer. Math. Soc., Vol. 22 (1921), pp. 327—482 concerning the projective embedding of abelian varieties.

We have mentioned the above works, and these examples, to emphasize our indebtedness to these classical geometers, and to make it clear that our omission of a more complete list of their works should not be interpreted as a lack of appreciation for them. A similar remark applies as well to more recent work dealing with the classical case, for instance (to mention only two): J. Igusa, "On the Picard variety attached to algebraic varieties," Amer. J. Math.,Vol. 74 (1952), pp. 1—22, and K. Kodaira, "Characteristic linear systems of complete continuous systems," Amer. J. Math., Vol. 78 (1956), pp. 716—744.

Let us say a few words concerning the more precise manner in which the theory presented in this book was born.

It stems from two great lines of thought, the arithmetic line and the geometric line. The first one begins with the Riemann hypothesis, translated for function fields of one variable by Artin in his thesis [2]. It was Hasse [33] who proved it first for curves of genus 1, by a method which clearly pointed out its connection with the theory of correspondences (Deuring).

Independently, Hilbert was the first to have the intuition that the zeros of the zeta function are proper values of an operator. But which one?

The geometric line is that of the Italian geometers, principally Castelnuovo and Severi, as we have said above, who developed the theory of abelian varieties and correspondences from the geometric (and essentially algebraic) point of view.

In an article as brief as it is striking [82], Weil unites these two lines and shows how the theory of correspondences is related with the Riemann hypothesis for curves by the theorem of

Castelnuovo-Severi on the equivalence defect, how one can define l-adic representations which give the equivalent of the representation of a correspondence in the first homology group, and how the zeta function (of a curve) is essentially the characteristic polynomial of the Frobenius transformation relative to these representations. The zeros were indeed characteristic roots of an operator, and by the same token, Weil also proved Artin's conjecture concerning the L-series.

There remained but to justify all of this by laying solid foundations for intersection theory. Contributions had been made on this problem previously by Severi, Van der Waerden, and simultaneously by Chevalley, but the date of publication of *Foundations of Algebraic Geometry* may be taken to be the beginning of the era that we consider below.

Let us note that the zeta function is still acting as a powerful motivator for both arithmetic and purely algebraic research, for instance through Weil's conjectures relating it to the Lefschetz fixed-point formula [86], which requires an algebraic definition of the homology groups. Other arithmetic questions, for instance class field theory [43], [47] give rise to problems in the theory of coverings, and complex multiplication [32], [77], [78], [80], [96] in the theory of algebraic systems of abelian varieties (of which the theory of reduction modulo p is a special case).

I have also included in the bibliography other papers dealing with arithmetic questions which go more deeply into the structure of the group of rational points of an abelian variety over special fields of interest to the arithmetician. These include p-adic fields [52], [62], [41], [81], finite fields [10], [11], [45], or number fields [50], [67], [68]. These papers represent only a sample of the available literature. The general diophantine problems, and their special formulations for abelian varieties, continue to represent some of the strongest aesthetic attractions of algebraic geometry.

[1] A. Albert, On involutorial algebras, Proc. Nat. Acad. Sci. U.S.A., Vol. 4, No. 7 (1955), pp. 480—482.

[2] E. Artin, Quadratische Körper im Gebiete der höheren Kongruenzen, I and II, Math. Zeit., 19 (1924), pp. 153—246.

[3] I. Barsotti, Structure theorems for group varieties, Annali di Matematica pura ed applicata, Serie IV, T. 38 (1955), pp. 77—119.

[4] ———, Un teorema di struttura per le varieta gruppali, Acad. Naz. dei Lincei, Serie VIII, Vol. 18 (1955), pp. 43—50.

[5] ———, Il teorema di dualita per le varieta abeliane ed altri resultati, Rend. di Mat. e delle sue applicazioni, Serie V, Vol. XIII (1954), pp. 1—17.

[6] ———, A note on abelian varieties, Rendiconti del Circolo Matematico di Palermo, Serie II, T. 2 (1954), pp. 1—22.

[7] ———, Abelian varieties over fields of positive characteristic, Rendiconti del circolo matematico di Palermo, Serie II, T. V. (1956), pp. 1—25.

[8] ———, Factor sets and differentials on abelian varieties, Trans. Am. Math. Soc., Vol. 84, No. 1 (1957), pp. 85—108.

[9] ———, Gli endomorfismi delle varieta abeliane su corpi di caratteristica positiva, Annali della Scuola Normale Superiore di Pisa, Serie III, Vol. X, Fasc. I—II (1956), pp. 1—23.

[10] F. Chatelet, Sur l'arithmétique des courbes de genre 1, Annales de l'université de Grenoble, Tome XXII (1946), pp. 153—165.

[11] ———, Les courbes de genre 1 dans un champs de Galois, Comptes rendus de l'Académie des Sciences, T. 224 (1947), pp. 1616—1618.

[12] ———, Methode Galoisienne et courbes de genre 1, Annales de l'université de Lyon, Section A, IX (1946), pp. 40—49.

[13] ———, Variations sur un thème de Poincaré, Annales Scientifiques de l'Ecole Normale Supérieure, Vol. 59 (1944), pp. 249—300.

[14] W. L. Chow, Abelian varieties over function fields, Trans. Amer. Math. Soc., Vol. 78 (1955) pp. 253—275.

[15] ———, Abstract theory of the Picard and Albanese varieties, to appear in the Annals of Math.

[16] ———, On the defining field of a divisor in an algebraic variety, Proc. Amer. Math. Soc., Vol. 1, No. 6 (1950), pp. 797—799.

[17] ———, The Jacobian variety of an algebraic curve, Amer. J. Math., Vol. 76, No. 2 (1954), pp. 453—476.

[18] ———, Projective embedding of homogeneous speces, Lefschetz conference volume, Princeton (1957).

[19] ———, On equivalence classes of cycles in an algebraic variety, Ann. Math., Vol. 64, No. 3 (1956), pp. 450—479.

[20] ———, On the quotient variety of an abelian variety, Proc. Nat. Acad. of Sciences U.S.A., Vol. 38 (1952), pp. 1039—1044.

[21] ———, On the principle of degeneration in algebraic geometry, Ann. Math., 66 (1957), pp. 70—79.

[22] W. L. Chow and S. Lang, On the birational equivalence of curves under specialization, Amer. J. Math., Vol. 79 (1957), pp. 649—652.

[23] M. Deuring, Die Typen der Multiplikatorenringe elliptischer Funktionenkörper, Abh. Math. Sem. Hams. Univ., 14 (1941), pp. 197—272.

[24] ———, Reduktion algebraischer Funktionenkörper nach Primdivisoren des Konstantenkörpers, Math. Zeit., 47 (1942), pp. 643—654.

[25] ———, Zur Theorie der elliptischen Funktionenkörper, Abh. Math. Sem. Univ. Hamburg, 15 (1947), pp. 211—261.

[26] ———, Algebraische Begrundung der komplexen Multiplikation, Abh. Math. Sem. Univ. Hamburg, 16 (1947), pp. 32—47.

[27] ———, Teilbarkeitseigenschaften der singularen Moduln der elliptischen Funktionen und die Diskriminante der Klassengleichung, Comm. Math. Helv., 19 (1947), pp. 74—82.

[28] ———, Die Anzahl der Typen von Maximalordnungen einer definiten Quaternionenalgebra mit primer Grundzahl, Deutschen Math. Ver., 54, 1 (1944), pp. 24—41.

[29] ———, Die Struktur der elliptischen Funktionenkörper und die Klassenkörper der imaginären quadratischen Zahlkörper, Math. Annalen, 124 (1952), pp. 393—426.

[30] ———, Die Zetafunktion einer algebraischen Kurve vom Geschlechte Eins, Drei Mitteilungen, Nachrichten der Akademie der Wissenschaften in Göttingen, 1953, pp. 85—94, 1955, pp. 13—43, and 1956, pp. 37—76.

[31] ———, Zur Transformationstheorie der elliptischen Funktionen, Akademie der Wissenschaften und der Literatur, Abh. Math. Naturwissenschaftlichen Klasse, Nr. 3 (1954), pp. 95—104.

[32] ———, On the zeta function of an elliptic function field with complex multiplication, Proceedings of the International Symposium on Algebraic Number Theory, Tokyo-Nikko, 1956, pp. 47—50.

[33] H. Hasse, Abstrakte Begründung der komplexen Multiplikation und Riemannsche Vermutung in Funktionenkörpern, Abh. Math. Sem. Univ. Hamburg, 10 (1934), pp. 325—348.

[34] ———, Zur Theorie der abstrakten elliptischen Funktionenkörper I, II, and III, J. Reine Angew. Math., 175 (1936).

[35] ———, Über die Riemannsche Vermutung in Funktionenkörpern, Comptes Rendus du congrès international des mathématiciens, Oslo, (1936), pp. 189—206.

[36] ———, Existenz separabler zyklischer unverzweigter Erweiterungskörper vom Primzahlgrade p über elliptischen Funktionenkörpern der Charakteristik p, J. Reine Angew. Math., 172, 2 (1934), pp. 77—85.

[37] J. Igusa, On some problems in abstract algebraic geometry, Proc. Nat. Acad. Sci. U.S.A., Vol. 41, No. 11 (1955), pp. 964—967.

[38] ———, A fundamental inequality in the theory of Picard varieties, Proc. Nat. Acad. Sci. U.S.A., Vol. 41, No. 5 (1955), pp. 317—320.

[39] ———, Fibre systems of Jacobian varieties, Amer. J. Math., Vol. 78, No. 1 (1956), pp. 171—199.

[40] ———, Fibre systems of Jacobian varieties part II (Local monodromy groups of fibre systems), Amer. J. Math., Vol. 78, No. 4 (1956), pp. 745—760.

[41] ———, Analytic groups over complete fields, Proc. Nat. Acad. of Sci. U.S.A., Vol. 42, No. 8 (1956), pp. 540—541.

[42] S. Lang, Abelian varieties over finite fields, Proc. Nat. Acad. Sci. U.S.A., Vol. 41, No. 3 (1955), pp. 174—176.

[43] ———, Unramified class field theory over function fields in several variables, Ann. Math., Vol. 64, No. 2 (1956), pp. 285—325.

[44] ———, On the Lefschetz principle, Ann. Math., Vol. 64, No. 2 (1956), pp. 326—327.

[45] ———, Algebraic groups over finite fields, Amer. J. Math., Vol. 78, No. 3 (1956), pp. 555—563.

[46] ———, L-series of a covering, Proc. Nat. Acad. Sci. U.S.A., Vol. 42, No. 7, (1956), pp. 422—424.

[47] ———, Sur les séries L d'une variété algébrique, Bull. Soc. Math. France, 84 (1956), pp. 385—407.

[48] ———, Divisors and endomorphisms on an abelian variety, Amer. J. Math., Vol. 79, No. 4 (1957), pp. 761—777.

[49] S. Lang and J.-P. Serre, Sur les revêtements non ramifiés des variétés algébriques, Amer. J. Math., Vol. 79, No. 2 (1957), pp. 319—330.

[50] S. Lang and J. Tate, Principal homogeneous spaces over Abelian varieties, Amer. J. Math., Vol. 80, No. 3 (1958), pp. 659—684.

[51] S. Lang and A. Weil, Number of points of varieties in finite fields, Amer. J. Math., Vol. 76, No. 4 (1954), pp. 819—827.

[52] E. Lutz, Sur l'équation $y^2 = x^3 - Ax - B$ dans les corps p-adiques, J. Reine Angew. Math., 177, 4 (1937), pp. 238—247.

[53] T. Matsusaka, On the algebraic construction of the Picard variety, I and II, Jap. J. Math., Vol. XXI (1951), pp. 217—235 and Vol. XXII (1952), pp. 51—62.

[54] ———, On algebraic families of positive divisors and their associated varieties, J. Math. Soc. Japan, Vol., 5, No. 2 (1953), pp. 113—136.

[55] ———, Specialization of cycles on a projective model, Memoirs of the College of Science, University of Kyoto, Series A, Vol. 26, No. 2 (1950), pp. 167—173.

[56] ———, Some theorems on abelian varieties, Natural Science Report, Ochanomizu University, Vol. 4, No. 1 (1953), pp. 22—35.

[57] ———, A remark on my paper Some theorems on abelian varieties, Natural Science Report, Ochanomizu University, Vol. 4, No. 2 (1953), pp. 172-174.

[58] ———, A note on my paper Some theorems on abelian varieties, Natural Science Report, Ochanomizu University, Vol. 5, No. 1 (1954), pp. 21—23.

[59] ———, On the theorem of Castelnuovo-Enriques, Natural Science Report, Ochanomizu University, Vol. 4, No. 2 (1953), pp. 164—171.

[60] ———, The criteria for algebraic equivalence and the torsion group, Amer. J. Math., Vol. 79, No. 1 (1957), pp. 53—66.

[61] ———, Polarized varieties, Amer. J. Math., Vol. 80, No. 1 (1958), pp. 45—82.

[62] A. Mattuck, Abelian varieties over p-adic fields, Ann. Math., 62 (1955), pp. 92—119.

[63] H. Morikawa, On abelian varieties, Nagoya Math. Journal, Vol. 6 (1953), pp. 151—170.

[64] ———, Cycles and endomorphisms on an abelian variety, Nagoya Math. Journal, Vol. 7 (1954), pp. 95—102.

[65] S. Nakano, On invariant differential forms on group varieties, M. Math. Soc. Japan, Vol. 2 (1951), pp. 216—227.

[66] ———, Note on group varieties, Mem. Coll. Sci. Univ. Kyoto, Vol. XXVII (1952), pp. 55—66.

[67] A. Néron, Problèmes arithmétiques et géométriques rattachés à la notion de rang d'une courbe algébrique dans un corps, Bull. Soc. Math. France, 80 (1952), pp. 101—166.

[68] ———, Arithmétique et classes de diviseurs sur les variétés algébriques, Proceedings of the International Symposium on Algebraic Number Theory, Tokyo-Nikko, (1955), pp. 139—154.

[69] A. Néron and P. Samuel, La variété de Picard d'une variété normale, Annales de l'institut Fourier, IV (1952), pp. 1—30.

[70] M. Rosenlicht, Equivalence relations on algebraic curves, Ann. Math., Vol. 56, No. 1 (1952), pp. 169—191.

[71] ———, Generalized Jacobian varieties, Ann. of Math., Vol. 59, No. 3 (1954), pp. 505—530.

[72] ———, Basic theorems on algebraic groups, Amer. J. Math., Vol. 78, No. 2 (1956), pp. 401—443.

[73] ———, Differentials of second kind for algebraic function fields of one variable, Ann. Math., Vol. 57, No. 3 (1953), pp. 517—523.

[74] ———, A universal mapping property of generalized Jacobians, Ann. Math., Vol. 66, No. 2 (1957), pp. 80—88.

[75] P. Samuel, Rational equivalence of arbitrary cycles, Amer. J. Math., Vol. 78, No. 2 (1956), pp. 383—400.

[76] J.-P. Serre, Sur la topologie des variétiés algébriques en caractéristique *p*. Congrès de topologie, Mexico, 1956,.

[77] G. Shimura, Reduction of algebraic varieties with respect to a discrete valuation of the basis field, Amer. J. Math., Vol. 77 (1955), pp. 134—176.

[78] ———, On complex multiplication, Proceedings of the International Symposium on Algebraic Number Theory, Tokyo-Nikko, 1955, pp. 23-30.

[79] E. Spanier, The homology of Kummer manifolds, Proc. Amer. Math. Soc., Vol. 7 (1956).

[80] Y. Taniyama, Jacobian varieties and number fields, Proceedings of the International Symposium on Algebraic Number Theory, Tokyo-Nikko, 1955, pp. 31—45.

[81] J. Tate, *WC*-groups over *p*-adic fields, to appear, Amer. J. Math.

[82] A. Weil, Sur les fonctions algébriques à corps de constantes fini, C. R. Acad. Sci. Paris, 210 (1940), pp. 592—594.

[83] ———, Foundations of algebraic geometry, Amer. Math. Soc. Colloquium Publications, Vol. XXIX, New York, 1946.

[84] ———, Sur les courbes algébriques et les variétés qui s'en deduisent, Hermann, Paris, 1948.

[85] ———, Variétés abéliennes et courbes algébriques, Hermann, Paris, 1948.

[86] ———, Number of solutions of equations in finite fields, Bull. Amer. Math. Soc., 55 (1949), pp. 497—508.

[87] ———, Arithmetic on algebraic varieties, Ann. Math., (2) 53 (1951), pp. 412—444.

[88] ———, Jacobi sums as Grossencharaktere, Trans. Amer. Math. Soc., 73 (1952), pp. 487—495.

[89] ———, Sur les critères d'équivalence en géométrie algébrique, Math. Annalen, 128 (1954), pp. 95—127.

[90] ———, On algebraic groups of transformations, Amer. J. Math., Vol. 77, No. 2 (1955), pp. 355—391.

[91] ———, Algebraic groups and homogeneous spaces, Amer. J. Math., Vol. 77, No. 3 (1955), pp. 493—512.

[92] ———, The field of definition of a variety, Amer. J. Math., Vol. 78, No. 3 (1956), pp. 509—524.

[93] ———, On the projective embedding of abelian varieties, in the volume in honor of S. Lefschetz, Princeton, 1957.

[94] ———, Zum Beweis des Torellischen Satzes, Nachrichten der Akad. Wiss. Göttingen, 1957, pp. 33—53.

[95] ———, On a certain type of characters of the ideal class group of an algebraic number field, Proceedings of the International Symposium on Algebraic Number Theory, Tokyo-Nikko, 1956, pp. 1—7.

[96] ———, On the theory of complex multiplication, *Ibid.*, pp. 10—22.

[97] E. Witt, Zyklische Körper und Algebren der Charakteristik p, vom grad p^m, J. Reine angew. Math., 176 (1936), pp. 126—140.

TABLE OF NOTATION

This is a brief list of notations used frequently throughout the book, together with a short informal description of each term.

X_a	:	Translate of a cycle by a point a on a group.
X^-	:	Image of a cycle by the map $u \to -u$ on a group.
$X * Y$	:	Pontrjagin product.
Θ	:	Divisor on a Jacobian, equal to the sum of the curve taken $g - 1$ times.
δ_A	:	Identity on an abelian variety A.
$D_a(U)$	:	Group of divisors algebraically equivalent to 0.
$D_l(U)$	:	Group of divisors linearly equivalent to 0.
$D_\tau(U)$	:	Torsion group relative to $D_a(U)$.
$\sim$	:	Linear equivalence.
$\approx$	:	Algebraic equivalence.
$\approxeq$	:	Numerical equivalence
$\mathrm{Cl}(X)$	:	Linear equivalence class of a divisor X.
φ_X	:	The homomorphism $u \to \mathrm{Cl}\,[X_u - X]$.
$\equiv$	:	The equivalence given by $X_u \sim X$ for all u.
$N_i(A)$	:	Group of i-cycles modulo numerical equivalence.
$N_*(A)$	:	Numerical equivalence ring of A.
$N'(A)$	:	Factor group of divisors by those which are $\equiv 0$.
κ_A	:	Canonical map of A onto $\hat{A}$. If D is a Poincaré divisor on $A \times \hat{A}$, it is $u \to \mathrm{Cl}\,[D(u) - D(0)]$.
$D_\xi(\alpha, \beta)$	:	The divisor class in $N'(A)$ given by $(\alpha + \beta)^{-1}(\xi) - \alpha^{-1}(\xi) - \beta^{-1}(\xi)$.
$D_\xi(\alpha)$	:	The divisor class $D_\xi(\alpha, \delta)$.
$\nu(\alpha)$	:	The degree of a map.
$Y \circ X$	:	The composed cycle $\mathrm{pr}_{13}\,[(X \times W) \cdot (U \times Y)]$ if X is on $U \times V$ and Y on $V \times W$.
$D \circ f$	:	Should be $D \circ \Gamma_f$.

A **Q** as a lower index means tensor product with **Q** (the rational numbers). As usual, **Z** are the integers.

INDEX

Abelian variety, II, p. 19
Abel's theorem, II, § 2, p. 36
Admissible map, II, § 3, p. 45
Albanese subfield, II, § 3, p. 45
Albanese variety, II, § 3, p. 41
Algebra of endomorphisms, II, § 1, p. 26
Algebraic group, I, § 1, p. 2
Algebraically equivalent to 0, III, § 1, p. 56
Ample linear system, IV, § 1, p. 87
Automorphism of polarized abelian variety, VII, § 2, p. 194

Birational isomorphism, I, § 1, p. 2

Canonical homomorphism on factor group, I, § 1, p. 3
Canonical map into Albanese variety, II, § 3, p. 41
Canonical map into Jacobian, II, § 2, p. 35
Characteristic polynomial, IV, § 3, p. 110
Characteristic roots, IV, § 3, p. 112
Chow's theorem, II, § 1, p. 26
Composed cycle, App., § 1, p. 231
Correspondence class, App., § 2, p. 240

Defined (composed cycle), App., § 1, p. 231
Defined over a field (rational map), I, § 1, p. 2
Degree, I, § 2, p. 12; IV, § 3, p. 103
Divisorial correspondence, App., § 2, p. 240
Duality hypothesis, VIII, § 4, p. 216

Endomorphism, II, § 1, p. 25
Extended Tate group, VII, § 1, p. 181

Fundamental theorem on symmetric functions, I, § 1, p. 6

Generate, II, § 3, p. 40
Generic homomorphism, I, § 1, p. 5
Generically exact, VIII, § 6, p. 224
Group variety, I § 1, p. 2

Homogeneous space, I, § 1, p. 3
Homomorphism (of an algebraic group into another), I, § 1, p. 2

Induced homomorphism into the Albanese variety, II, § 3, p. 41
Inverse of homomorphism, II, § 1, p. 29
Isogenous, II, § 1, p. 29
Isomorphism, I, § 1, p. 2

Jacobian, II, § 2, p. 33

k-Albanese variety, II, § 3, p. 45
K/*k*-image, VIII, § 1, p. 199
K/*k*-maximal abelian subvariety, VIII, § 3, p. 212
K/*k*-trace, VIII, § 3, p. 213

Left translation, I, § 1, p. 2
Linear conditions at a simple point, III, § 4, p. 77
Linearly equivalent to 0, III, § 1, p. 64

Néron-Severi group, III, § 1, p. 64
Non-degenerate divisor, IV, p. 85

Normal law of composition, I, § 1, p. 4
Numerically equivalent to 0, IV, § 3, p. 101

Parameter variety, III, § 1, p. 56
Picard group, III, § 1, p. 64
Picard variety, IV, § 4, p. 114
Poincaré divisor, IV, § 4, p. 114
Polar class, IV, § 3, p. 112
Polar divisor, IV, § 3, p. 108
Polarized abelian variety, VII, § 2, p. 194
Pontrjagin product, I, § 2, p. 8
Positive endomorphism, V, § 3, p. 141
Principal homogeneous space, I, § 1, p. 4
Pure variety, II, § 1, p. 25

Rational homomorphism, I, § 1, p. 2
Reflexive abelian variety, V, § 2, p. 129
Regular exact sequence, VIII, § 4, p. 216
Regularity theorem, VIII, § 1, pp. 204, 214

Seesaw principle, App., § 2, p. 241
Semi-pure variety, II, § 1, p. 25
Simple abelian variety, II, § 1, p. 29
Squarely equivalent to 0, III, § 3, p. 75
Sum, I, § 2, p. 7
Symmetric endomorphism, V, § 3, p. 141

Tate group, VII, § 1, p. 180
Theorem of the cube, III, § 2, p. 68
Theorem of the square, III, § 2, p. 68
Torsion divisor, III, § 1, p. 65
Torsion group, III, § 1, p. 65
Trace, IV, § 3, p. 110
Transfer (of normal law of composition), I, § 1, p .4
Translation, I, § 1, p. 2
Transpose of homomorphism, V, § 1, p. 124
Trivial correspondence or divisor, App., § 2, p. 240

Zeta function of a variety, V, § 4, p. 139

9 780387 908755